OXFORD STUDIES IN PHYSICS

GENERAL EDITORS

B. BLEANEY, D. W. SCIAMA, D. H. WILKINSON

COSMIC
DUST
ITS IMPACT ON ASTRONOMY

BY

PETER G. MARTIN

1978

CLARENDON PRESS · OXFORD

Oxford University Press, Walton Street, Oxford OX2 6DP

OXFORD LONDON GLASGOW
NEW YORK TORONTO MELBOURNE WELLINGTON
IBADAN NAIROBI DAR ES SALAAM LUSAKA CAPE TOWN
KUALA LUMPUR SINGAPORE JAKARTA HONG KONG TOKYO
DELHI BOMBAY CALCUTTA MADRAS KARACHI

ISBN   0   19   851458   1

© Oxford University Press 1978

Printed in Great Britain

by Thomson Litho Ltd,

East Kilbride

TO
LIZ

PREFACE

'One problem I have in mind would be to make a study of dust
particles in our own Galaxy and other galaxies. I believe this
subject has a great future. This [is a] problem which despite
its messiness is of great astrophysical importance.'

In this way D.W. Sciama guided me towards this fascinating field in 1969.
At this time I would pass on much the same recommendation. The subject is
indeed enjoying 'a great future', in the interim being spurred on in large
part by new observational data gathered at the heels of various technologi-
cal advances; at the same time it has become no less messy! In fact we
find ourselves in the midst of a revolution in many of our ideas, keeping
pace with similar exciting discoveries relating to the interstellar medium.
The breadth of topics to be addressed in this book serves to re-emphasize
the importance of dust to the understanding of diverse astrophysical phenomena.

Preparing a complete and up-to-date summary of an actively developing
subject is somewhat precarious. Therefore in this introduction to interstellar dust and its impact on astronomy I have concentrated on providing
the tools and discussing the concepts which underlie current research,
though not at the expense of denying the speculative and ephemeral aspects.
Interwoven in this development are informed statements of our present
understanding and some suggestions for the future.

The organization of this book is intentionally somewhat pedagogical.
Following a brief historical introduction and overview of current observational imperatives in Chapter 1, six chapters are devoted to the three fundamental ways in which dust is studied, namely through its effects on the
transmission of radiation in the interstellar medium, its somewhat complementary role in scattering, and its thermal emission. Each pair of chapters commences with an exposition of the basic concepts and lays the theoretical basis by which observational tests are conceived and data are interpreted. Specific applications are then treated in further detail to conso-
lidate this rudimentary material. A discussion of various surface phenomena and important dynamical questions ensues. This introduction is completed by a wide-ranging review of the occurence of dust throughout our
Galaxy.

Determining the composition of the dust particles is the most
intriguing and fundamental problem, on whose solution rests the interpretation of all interactions involving the grains. Arguments bearing on the

composition are marshalled in Chapter 10, followed in Chapter 11 by a discussion of the processes which can create and destroy such grains. A fitting close to this book is a glimpse at one exciting frontier area of
research: on dust in other galaxies and in intergalactic space.

The literature cited at the end of each chapter and in some figure
captions is not intended to serve the purpose of the bibliography of a
review article (several such reviews are mentioned in Chapter 1). While
this literature is current and does address many specific issues raised in
the text, it is to be regarded chiefly as an introduction to the vast
research literature from which one can trace the roots or branch forth into
wider spaces.

Before becoming engrossed in our topic we might all be well advised
to consider a longer view, such as that expressed by C. Schalén a decade
ago (Astron. Nachr.   293, 1 (1971)):

'But is it astonishing that we have not yet definite conclusions concerning interstellar dust? I do not believe so,
because, after all, this is a very young branch of astronomy.
One cannot expect to have these intricate problems solved in
forty years.'!

Toronto, July 1978   P.G.M.

ACKNOWLEDGEMENTS

Without the assistance and co-operation of many people this enterprise would surely have floundered. I am greatly indebted to J. Yamazaki both for typing the manuscript file and for adding the myriad of special symbols to the camera-ready typescript that I formatted using the text processor SCRIPT. D. Grazioli and P. Sullivan typed earlier drafts of several chapters. The figures were produced mainly in the Graphics Department of Scarborough College in the University of Toronto. T. Westbrook prepared the line drawings, D. Harford undertook the photographic copying, and L. McGregor composed the tables. Some computer-produced diagrams were developed by I. Prociuk and C. Rogers. Original photographs were provided by D.E. Brownlee, M.P. Fitzgerald, D.A. Harper, C.J. Lada, G.D. Schmidt, J.W. Tapscott, and A.N. Witt. K. Kamper of the David Dunlap Observatory reproduced the other photographs. Figures used with permission from various authors, observatories, and publishers are acknowledged individually in the appropriate captions.

J. Percy provided generous access to his collection of astronomical journals. Timely communications from B. Balick, W.W. Duley, J.D. Fernie, R.C. Henry, A.R. Holt, D.A. Huffman, J.S. Mathis, D.N. Schramm, N.C. Wickramasinghe, and D.G. York were appreciated. The manuscript has benefitted from editorial scrutiny by my publisher and the comments by their scientific advisor. C. Rogers suggested many improvements, as did E.M. Purcell who also provided unpublished material concerning grain alignment. To each of the above I extend my thanks and to those adversely affected by my devotion to this project I offer my apologies.

My research on interstellar dust is supported by the Natural Sciences and Engineering Research Council of Canada.

CONTENTS

1

INTRODUCTION

1.1. DUST IN TIME

If, as we contemplate our origins, our curiosity were allowed to lead us
back sufficiently far, we would undoubtedly discover many ancestral interstellar dust particles. These small particles contain most of the material, other than H and He, in the interstellar medium, and it is from such
elements that the terrestrial planets and we ourselves have formed. Consideration of the relevant time scales - beginning with the origin of H and
He in the 'big bang', progressing to formation of our Galaxy and subsequent
nucleosynthesis of heavier elements in stars, through several generations
of interstellar particles to the origin of the solar system and our evolution - shows that our constituent mass must have spent a significant fraction of time, perhaps as much as half, condensed as interstellar dust.

By our standards interstellar space is sparsely populated; for example each of us, when distributed as small particles at average interstellar
densities, would have occupied about 5×10^{30} cm^3 (about 5000 times the
volume of the earth). By similar standards the life of an interstellar
dust grain is lonely; it is visited by a H atom only a few times a day.

If we pursue this fanciful anthropomorphic view further, on a Galactic time scale, then we find that the life of a grain in not uneventful.
Consider one possible scenario. A particle is born in the warm gas surrounding a star and is blown into interstellar space. When it has cooled
off it grows a coating of more volatile elements. Such a mantle is somewhat ephemeral, being shed in a variety of circumstances including the formation of a hot star nearby, and the passage of an interstellar shock; however, it is soon grown again after conditions return to normal.

A dust particle faces a number of crises which can prove fatal. For
example if caught in the worst of the shocks, from a supernova explosion, a
grain usually succumbs. Regions of star formation are also a menace.
After a brief interlude in which a new coating is acquired, a grain, quite
betrayed, could find itself totally vaporized. From this pyre, however, a
new grain can arise like the phoenix. Renewal can be quite prompt if the
material escapes being swallowed up by the star, but otherwise there is
considerable (if not indefinite) delay until the star experiences mass
loss, perhaps as a supernova, planetary nebula, or cool stellar wind. We
ourselves are such an interruption in the cycling of interstellar dust, but
our days are numbered; we can expect to be returned to the dusty realm when
our sun becomes a red giant, 5×10^9 y hence.

1.2. AN EDIFICE OF DUST

We shall see that quite a grand structure has already been erected on the
basis of observations and theories of interstellar dust. Lately there have
been many additions, much partitioning, and some embellishments. Although
construction continues apace, substantial renovations, if not major gutting
and rebuilding, can be anticipated before the edifice can be declared complete.

        Our discussion will deal mostly with the development of the blue-
prints, and their execution and interpretation, so that the readers,
through this apprenticeship, can become the artisans of the future. First,
however, let us briefly examine our heritage.

1.2.1. Early excavations

During his 'star sweeps' of 1784, W. Herschel, noticing a region of sky
particularly devoid of stars, was prompted to exclaim 'Hier ist wahrhaftig
ein Loch im Himmel'. (This region can be seen in Fig. 9.4 of the ρ Oph
complex, near the centre of the photograph just to the east of the globular
cluster M 80. Note that a visual observer would not detect the reflection
nebulosity.) Although Herschel also noted the spatial correlation of clusters, bright nebulae, and these 'dark holes', he interpreted this as evidence for the breaking up of the Milky Way. Proper recognition of these
'dark holes' as silhouettes of obscuring dust clouds had to await the new
technology of celestial photography, which as Barnard remarked in 1889
'showed for the first time in all their delicacy and beauty the vast and
wonderful cloud-forms, with their remarkable structure of lanes, holes, and
black gaps, and sprays of stars, as no eye or telescope can even hope to
see them' (see Fig. 9.5). Barnard's Atlas of selected regions of the Milky
Way was published in 1927, the culmination of 30 years of investigation.
In the meantime Curtis was accumulating evidence for obscuring material in
spiral nebulae, particularly evident as dark equatorial bands (see Fig.
12.1). By supposing that the Milky Way was a spiral nebula he reasoned
that similar obscuration in the plane of our Galaxy (see Fig. 9.1 for a
modern montage) could explain the 'zone of avoidance', i.e. the absence of
globular clusters and spiral nebulae in the Milky Way. A relationship between Barnard's dark clouds and this phenomenon was suggested. (Wolf made a
similar comparison with Andromeda in 1908.) Russell too favoured a flattened dark-cloud distribution as an explanation of the different spatial
distributions of globular and open clusters. However, in 1920 these views
were not generally accepted.

        Reflection nebulae were also under study. Slipher discovered in 1912
that the spectrum of the nebulosity in the Pleiades (Fig. 5.1) was the same

as that of the illuminating stars, in contrast to the bright-line spectrum found previously for many other diffuse nebulae (H II regions). By 1919 the ρ Oph nebula (Fig. 9.4) and five others had been added to the list of reflection nebulae. Somewhat later (1922), after Hubble had established the relationships of type and angular size of diffuse nebulae to the spectral type and apparent brightness of the associated stars, Russell emphasized the similarities of dark clouds and reflection nebulae, and argued for millimetre-sized dust particles. Pannekoek's studies in 1920 of the Taurus dark clouds also suggested a small size for the particles, comparable with the wavelength of visible light. This particular choice has the desired effect of minimizing the inferred mass of the dark cloud, since all alternatives - Rayleigh scattering by atoms and molecules, electron scattering, and extinction by larger meteoritic grains - are less efficient in terms of extinction per unit mass.

The patchy appearance of dark clouds, together with their sharp edges, left open the possibility that the intervening interstellar space was transparent. Kapteyn, among those using star counts to develop models of the 'universe', was aware that undetected general obscuration would make stellar distance estimates too large and would cause an apparent decrease in stellar densities with distance. However, no widely accepted demonstration of extinction, or of reddening that would arise from wavelength-dependent extinction, was available, although the issue had been addressed much earlier (Chéseaux in 1740, Olbers in 1823, Struve in 1847) and new attempts using photographic data had been made. Thus the question of general obscuration was to remain, in the words of Eddington, 'a bogey which threatened the security of many of our theories of the structure and mechanism of the stellar universe'. (Indeed the effects of extinction introduce an element of uncertainty and controversy even today, as exemplified by discussions of possible variations in the ratio of total to selective extinction.) Shapley's evidence for the lack of reddening of distant objects, first globular clusters (in 1917) and later spiral nebulae (in 1926), seems to have been persuasive; what he overlooked in his pronouncements on the transparency of space was the possibility that diffuse obscuring material was concentrated in the plane of the Milky Way, an effect already suggested by the distribution of dark clouds and the existence of the 'zone of avoidance'.

1.2.2. The foundations

A resolution of conflicting opinions and synthesis of varied observational data became possible after 1930, when Trumpler presented convincing evidence for a thin absorbing layer in the Galactic plane. His technique was to compare luminosity distances to open clusters, as derived from their Hertzsprung-Russell diagrams, with distances judged from their angular

diameters; the former estimates are sensitive to extinction whereas the
latter, being geometrical, are not. Trumpler also used the increase in
average colour excess with distance to demonstrate interstellar reddening.
(Essentially the same method, but with photoelectric data, has been used
recently to derive a ratio of total to selective extinction.)

Soon after, van de Kamp showed that the 'zone of avoidance' for globular clusters was consistent with a thin Galactic absorbing layer; in 1934
Hubble used the same explanation for the apparent distribution of spiral
galaxies. Stebbins managed to measure colour excesses of globular clusters
photoelectrically in 1933, finding the largest values near the 'zone of
avoidance'. By 1932 an upper limit (the 'Oort limit') to the local density, including stars and gas as well as dust, placed severe restrictions
on the source of the obscuration; only with small dust particles could so
much extinction by so little mass be explained. Some early interpretations
concentrated on metallic particles, based on an analogy with micrometeor-
ites supposedly broken down into finer dust.

1.2.3. The framework

Among the important advances made possible by photoelectric photometry was
an accurate determination of the wavelength dependence of extinction. By
1939 a λ^{-1} dependence in the optical region (rather than λ^{-4} appropriate to
Rayleigh scattering) was established; the definitive six-passband extinc-
tion curve published by Stebbins and Whitford in 1943 demonstrated curvature in the infrared (1.03 μ) and ultraviolet (0.35 μ) regions. Scattering
in reflection nebulae and diffuse Galactic light was interpreted by Henyey
and Greenstein in 1941 as evidence for forward-scattering grains with a
high albedo.

A new interpretation of the particle composition originated with
Lindblad in 1935, who considered the possibility of nucleation and growth
of icy dielectric grains in the interstellar medium. Oort and van de Hulst
computed an equilibrium size distribution, balancing growth by accretion
with destruction by mutual collisions during encounters of interstellar
clouds. Despite difficulties with the origin of the nucleating particles,
this hypothesis, particularly as presented by van de Hulst in 1949, became
widely accepted.

The scope of astrophysical investigations widened to include such
diverse topics as radiation pressure, equilibrium grain temperature, and
electric potential.

The serendipitous discovery of interstellar linear polarization in
1948 by Hall and Hiltner during searches for intrinsic polarization in early-type stellar atmospheres heralded a new and fruitful mode of investigation. Grain alignment with respect to an interstellar magnetic field was

postulated; the paramagnetic relaxation mechanism put forward by Davis and
Greenstein in 1951 seemed particularly attractive. Throughout that decade
major surveys of polarization throughout the Galactic plane were completed.
A different program, to measure the wavelength dependence of polarization
and thereby characterize the grains, was begun by Gehrels around 1960 and
continued by Coyne and Serkowski. Systematic polarization measurements of
reflection nebulae were begun by Elvius and Hall in 1962.

In one attempt to explain the degree of polarization Cayrel and
Schatzman proposed a new grain material, graphite. Hoyle and Wickramasinghe later (1962) suggested that graphite could condense in the atmospheres of cool carbon stars; the possibility of ejection of such circumstellar particles into the interstellar medium introduces an important
mechanism for grain formation.

The challenge presented by the new extinction, polarization, and
scattering data was met by more detailed modelling based on cross-sections
calculated according to the Mie theory for spherical particles and its
equivalent for infinite circular cylinders. (Computations for other shapes
have become possible only recently.) These potentially laborious computations were facilitated by the concurrent development of high-speed compu-
ters; in fact testing of the effects of various composition and size distributions became possible. Inhomogeneous particles, with a refractory
core and a more volatile mantle, were also considered.

1.2.4. Recent additions and alterations

Before a consensus could emerge, spectacular observational advances forced
a re-examination of many basic facets of the hypotheses that had been constructed. Some, like the measurement of the wavelength dependence of newly
discovered interstellar birefringence, seem to eliminate certain types of
materials such as graphite as sources of linear polarization. Others, like
the detection of strongly increasing extinction in the far ultraviolet,
require a new component of interstellar grains, in addition to those tradi-
tionally held responsible for the familiar extinction curve at optical wavelengths.

A major driving force behind these advances is new technology, which
opens new frontiers to exploration just as photography did for Barnard.
Infrared astronomy, carried out first at ground-based observatories and
then from balloons and high-flying aircraft, led to the detection of thermal emission from warm dust grains. This fruitful area of research continues to turn up spectral features that help identify the grain materials,
allows investigation of regions of star formation, and provides insight
into the condensation of grains in cool stellar envelopes. Ultraviolet
astronomy has blossomed too, benefiting from rocket-borne and then satel-

lite-borne telescopes. The principal spectroscopic discovery is a ubiquitous extinction bump at 2200 Å. Ultraviolet observations of scattered
radiation, as diffuse Galactic light and reflection nebulae, are also valuable.

These findings encourage a much fuller analysis of interstellar dust,
the rudiments of which are to be described below. However, it would be
misleading to pretend that we have arrived at the stage of decorating our
completed edifice. Many of the present studies, still in preliminary
stages, look forward to continued expansion, and better models are needed
to extract quantitative information from various data that are now understood only qualitatively. Furthermore the observational resources are by
no means exhausted. Measurements of X-ray scattering by grains, ultraviolet spectropolarimetry, more complete wavelength coverage in infrared
spectroscopy, and far-infrared photometry are among the developments that
can be anticipated for the near future. Theoretical and laboratory studies
will have to scramble to keep pace! We can hope that such an expanded data
base, and its interpretation and integration into theoretical investigations, will be the pawl and ratchet of a secure understanding of interstellar dust.

A final important point not stressed so far, but one apparent on even
cursory reflection, is the symbiotic relationship between research on dust
and on gas in the interstellar environment. Two significant early studies
were of interstellar absorption lines from diffuse gas and of the nature of
H II regions; two recent developments with considerable impact concern
interstellar molecules and the discovery of widespread depletion of heavy
elements in the gas phase. Such parallel and interacting investigations
will be important in our exposition, since it is through this dynamic that
dust becomes important to the advancement of astronomy on a broad front.

FURTHER READING

§1.2.1

CLERKE, A.M. (1901). The Herschels and modern astronomy. Cassel, London.
WATERFIELD, R.L. (1938). A hundred years of astronomy. Duckworth, London.
SEELEY, D., and BERENDZEN, R. (1972). The development of research in
        interstellar absorption, c. 1900-1930. J. Hist. Astron. 3, 52, 75.

§1.2.3

DUFAY, J. (1957). Galactic nebulae and interstellar matter. Hutchinson,
        London.
Les Particules Solides dans les Astres, 1954. M. Soc. r. Sci. Liege 15, 1
        (1955).

STRUVE, O., and ZEBERGS, V. (1962). Astronomy of the 20th century, Macmil-
    lan, New York.

§1.2.4

GREENBERG, J.M., and ROARK, T.P. (eds.) (1967). Interstellar grains, NASA
    Spec. Publ. SP-10. U.S. Government Printing Office, Washington, D.C.
WICKRAMASINGHE, N.C. (1967). Interstellar grains, Chapman and Hall, Lon-
    don.
GREENBERG, J.M. (1968). Interstellar grains. In Nebulae and interstellar
    matter (eds. B.M. Middlehurst and L.H. Aller). University of Chicago
    Press, Chicago.
LYNDS, B.T., and WICKRAMASINGHE, N.C. (1968). Interstellar dust. A. Rev.
    Astr. Astrophys. 6, 215.
ROSENBERG. J. (ed.) (1969). Int. Conf. on Laboratory Astrophysics. Phy-
    sica 41, 1.
IAU Colloq. on Interstellar Dust, 1969. Astron. Nachr. 293, 1 (1971).
WICKRAMASINGHE, N.C., and NANDY, K. (1972). Recent work on interstellar
    grains. Rep. Prog. Phys. 35, 157.
GREENBERG. J.M., and VAN DE HULST, H.C. (eds.) (1973). Interstellar dust
    and related topics. Reidel, Dordrecht.
AANNESTAD, P.A., and PURCELL, E.M. (1973). Interstellar grains. A. Rev.
    Astr. Astrophys. 11, 309.
GEHRELS, T. (ed.) (1974). Planets, stars and nebulae studied with photopo-
    larimetry. University of Arizona Press, Tucson.
FIELD, G.B., and CAMERON, A.G.W. (eds.) (1975). The dusty universe. Neale
    Watson Academic Publications, New York.
WICKRAMASINGHE, N.C., and MORGAN, D.J. (eds.) (1976). Solid state astro-
    physics. Reidel, Dordrecht.
ANDRIESSE, C.D. (1977). Radiating cosmic dust. Vistas Astron. 21, 112.
HUFFMAN, D.R. (1977). Interstellar grains: the interaction of light with
    a small-particle system. Adv. Phys. 26, 129.
SALPETER, E.E. (1977). Formation and destruction of dust grains. A. Rev.
    Astr. Astrophys. 15, 267.
McDONNELL, J.A.M. (ed.) (1978). Cosmic dust. Wiley, New York.
SPITZER, L. (1978). Physical processes in the interstellar medium. Wiley,
    New York.

2

TRANSMISSION OF RADIATION

2.1. THE EQUATION OF TRANSFER

2.1.1. Extinction and phase retardation

Dust particles have two distinct effects on the transmission of electromagnetic radiation (light): they cause a reduction in intensity, or extinction of light, and they introduce a phase shift. The extinction referred
to here includes both the effects of actual absorption within the particle
and of scattering radiation away from the line of sight. Individual attention is given to scattering and to absorption in Chapter 4 and Chapter 6,
respectively.

Extinction and phase retardation may be attributed formally to a complex refractive index $\tilde{m}$ of the interstellar medium, which being close to 1
(interstellar space is a near vacuum) will be written

$$\tilde{m} = 1 + (\epsilon - i\sigma)/k \tag{2.1}$$

where $k = 2\pi/\lambda$ is the wavenumber of the light and $i^2 = -1$. Twice σ is the
extinction per unit pathlength, whereas ϵ is the corresponding phase retardation. The factor A by which the complex amplitude of an incoming plane
wave is changed is

$$A = \exp\{-ik\ (\tilde{m}-1)\ ds\} \approx 1 - (\sigma + i\epsilon)\ ds \ . \tag{2.2}$$

A correspondence between σ and ϵ and the scattering properties of an individual particle can be found by calculating A independently for the latter
case.

The treatment of transmission, which deals with forward scattering
where the incident plane wave and scattered wave of the same polarization
are superimposed, requires an addition of amplitudes. Often the amplitude
E_s of the scattered wave is expressed in the form

$$E_s = -i\ \{\exp(-ikr)/kr\}\ S(0)\ E \tag{2.3}$$

where the factors r^{-1} (which gives the inverse square law) and k (which
makes S dimensionless) are excluded explicitly from the complex amplitude
function S. The zero indicates forward scattering. There is no unanimity
in the choice of sign in the exponential. Included in the convention
adopted here are a time dependence $\exp(-i\omega t)$ and $m = n - in$, $n' \geqslant 0$.

We are interested in $E' = E_s + E$ in an experimental situation in which E' is averaged either over a large aperture at a large distance from a particle or over a random arrangement of many particles adjacent to the line of sight. In either case

$$A = 1 - 2\pi k^{-2} S(0) n_d \, ds \qquad\qquad (2.4)$$

where n_d is the number density of dust particles. Clearly contributions from different types of particle are additive. $S(0)$ can be determined for individual particles, as described below.

The cross-section C_e for extinction, the area of the incident plane wave that would have to be completely blocked to account for the reduced intensity, is the quantity usually calculated and is related to the others by the optical theorem, which with our sign conventions is

$$C_e = 4\pi k^{-2} \, Re\{S(0)\} = 2\sigma/n_d \, . \qquad\qquad (2.5)$$

Similarly a phase-lag cross-section can be defined:

$$C_p = 4\pi k^{-2} \, Im\{S(0)\} = 2\varepsilon/n_d \, . \qquad\qquad (2.6)$$

It can be seen that the quantities σ and ε which are needed to study the transfer of radiation in the interstellar medium may be taken as the added effects of many individual particles for which the C's can be calculated. Often efficiency factors, called Q's, are given instead of C's; the latter are simply dimensionless numbers obtained by dividing the C's by some measure of the geometrical cross-section of the particles. For example, for a sphere of radius a, $Q_e = C_e/\pi a^2$ is the efficiency factor for extinction. When non-spherical particles are being considered it is often convenient to choose the area for normalization to be independent of orientation, so that the dependence of the C's on orientation can be seen directly from the Q's.

2.1.2. Polarization properties

Underlying our discussion of polarization properties is a picture of aspherical particles spinning so that on average their profile projected on a plane perpendicular to the line of sight is elongated with bilateral symmetry (Fig. 2.1), independent of the peculiar shapes of the grains. (This phenomenon, called grain alignment, is described further in §8.4.) Two principal directions, along the long and short directions, can be labelled I and II respectively, with the direction of propagation, axis III, completing the triad. (A somewhat different terminology will be applied to a

stationary particle.) Within a thin slice of the interstellar medium in
which the alignment is uniform, axis II of this profile-based co-ordinate
system will be fixed at some angle, say ψ, relative to a sky-based co-ordinate system, as depicted in Fig. 2.1.

To study the propagation in terms of two linearly polarized waves it
is convenient to take advantage of the symmetry and choose axes I and II as
the directions of vibration of the electric vectors, for then

$$\begin{pmatrix} E_I \\ E_{II} \end{pmatrix}' = \begin{pmatrix} A_I & 0 \\ 0 & A_{II} \end{pmatrix} \begin{pmatrix} E_I \\ E_{II} \end{pmatrix} .$$

(2.7)

All of the preceding discussion can be carried over by assigning subscripts
I and II to $\tilde{m}$, σ, ε, and C. In particular, C_I and C_{II} are the complex
extinction cross-sections when the electric vector lies along axes I and II
respectively. The linear extinction coefficient for natural light is $\delta =
\sigma_I + \sigma_{II}$. Two important properties, linear dichroism (differential extinc-
tion for the two linearly polarized waves) and linear birefringence (dif-
ferential phase shift), are described by $\Delta\sigma = \sigma_I - \sigma_{II}$ and $\Delta\varepsilon = \varepsilon_I - \varepsilon_{II}$
respectively. The absence of off-diagonal terms with this choice of axes
signifies the lack of both circular dichroism and birefringence.

The extinction and polarization properties of this medium can be analysed in terms of its effects on the Stokes vector, denoted $\underline{W}$ here. The
four components $\dot{I}$, Q, U, and V, which completely characterize the state of
the radiation field, are derived from the complex amplitudes of two waves

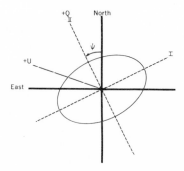

FIG. 2.1. Schematic representation of the mean grain profile
in a plane perpendicular to the line of sight. Axes I, II, and
the direction of propagation (III, directly out of the page)
form a right-handed system. The position angle ψ is measured
from north to east. The choice of directions for positive Q
and U, shown for the profile-based co-ordinate system, is consistent with this sense of measuring ψ.

having orthogonal linear polarization (and <u>vice</u> <u>versa</u>). Here we choose
positive Q to be the linearly polarized component along axis II, positive U
to be the linear component at $45°$ position angle (counter-clockwise) relative to axis II, and positive V to be the right-handed circularly polarized
component (a counter-clockwise rotation of the electric vector in the plane
perpendicular to the line of sight).

Formally we may write

$$\underline{W}' = (\underline{1} + \underline{\underline{H}} \ ds) \ \underline{W} , \quad \text{or} \quad d\underline{W}/ds = \underline{\underline{H}} \ \underline{W} . \qquad (2.8)$$

The matrix $\underline{\underline{H}}$ (seen in eqn (2.10) by putting $\psi=0$) is derived from eqn (2.7),
the definition of $\underline{W}$ in terms of the E's, and eqns (2.4)-(2.6) for the A's,
σ's, and ε's. Since each slice might have a different alignment ψ, it is
necessary when considering the whole medium to refer to a sky-based co-or-
dinate system through a rotation matrix $\underline{\underline{R}}$ which leaves I and V unchanged.
Choosing $\underline{W}_{sky} = \underline{\underline{R}} \ \underline{W}$ we have

$$d\underline{W}_{sky}/ds = \underline{\underline{H}}_{sky} \ \underline{W}_{sky} \qquad (2.9)$$

where

$$\underline{\underline{H}}_{sky} = \underline{\underline{R}} \ \underline{\underline{H}} \ \underline{\underline{R}}^{-1} = \begin{pmatrix} -\delta & \Delta\sigma \cos 2\psi & \Delta\sigma \sin 2\psi & 0 \\ \Delta\sigma \cos 2\psi & -\delta & 0 & \Delta\varepsilon \sin 2\psi \\ \Delta\sigma \sin 2\psi & 0 & -\delta & -\Delta\varepsilon \cos 2\psi \\ 0 & -\Delta\varepsilon \sin 2\psi & \Delta\varepsilon \cos 2\psi & -\delta \end{pmatrix}. \qquad (2.10)$$

Eqn (2.10) could also be reached by introducing the rotation ψ in eqn
(2.7).

Eqns (2.9) and (2.10) can be recast into equations for normalized
Stokes parameters, q(= Q/I), u, and v, which are especially simple when the
degree of polarization is small (< 20 per cent) and second-order and higher
terms may be dropped. These are (still for the sky-based co-ordinate system)

$$\begin{aligned} I^{-1} \ dI/ds &= -\delta , \\ dq/ds &= \Delta\sigma \cos 2\psi , \\ du/ds &= \Delta\sigma \sin 2\psi , \\ dv/ds &= \Delta\varepsilon \ (u \cos 2\psi - q \sin 2\psi) . \end{aligned} \qquad (2.11)$$

The production of extinction as measured by δ, the growth in linear polarization as measured by $\Delta\sigma$, and the process of linear-to-circular conversion
(as in a waveplate) are particularly clear in this formulation. Note that

the production of circular polarization is a second-order effect which
depends on either ψ changing along the line of sight or the presence of
suitably oriented intrinsic linear polarization in the source being
observed.

2.1.3. Uniform-slab models

Let us define a uniform slab to be a slice of the interstellar medium in
which ψ is constant. Within a single slab the four transfer equations (eqn
(2.11)) are easily integrated since the first three are uncoupled. A second instructive case, the combination of two uniform slabs, will also be
considered. It will be assumed that the incident radiation is unpolarized.

A single slab. The first familiar relationship introduces the optical depth τ:

$$I' = \exp(-\tau)\, I\ ; \quad \tau = \int \delta\ ds\ . \tag{2.12}$$

In the usual magnitude system of optical astronomy the extinction in magni-
tudes is simply $A(\lambda) \simeq 1.086\ \tau(\lambda)$. Often λ denotes a passband in some photometric system, like U, B, and V, and is written as a subscript. A change
in colour resulting from differential extinction can be measured by a colour excess $E(\lambda_1 - \lambda_2) = A(\lambda_1) - A(\lambda_2)$.

(Perhaps the most frequently encountered application is to photometric determinations of astronomical distances in which the apparent distance modulus has to be corrected for the effects of extinction. For a
source whose intrinsic colour is known, a colour excess is more readily
measured than the total extinction. Therefore, in the UBV system, A_V is
usually determined from the ratio of total to selective extinction,

$$R = A_V/E_{B-V}\ . \tag{2.13}$$

The object has been to calibrate R using more detailed observations of few
sources, and then to use this value for all other sources for which only
E_{B-V} is known (§3.1).)

The polarization produced by a uniform slab is described by

$$q = p \cos 2\psi, \quad u = p \sin 2\psi, \quad v = 0 \tag{2.14}$$

where

$$p = (q^2 + u^2)^{\frac{1}{2}} = \int \Delta\sigma\ ds\ . \tag{2.15}$$

Many early observations of linear polarization were reported in magnitude units, reflecting the basic technique of comparing magnitudes measured with orthogonal orientations of a polarization analyzer. This has now been superseded by the more conventional definition embodied by the Stokes parameters. The conversion formula is

$$p = \tanh \{0.4605 \, p(\text{mag})\} \approx 0.4605 \, p(\text{mag}) \, . \qquad (2.16)$$

For most applications in the interstellar medium the approximate version will suffice.

The position angle of the electric vector is defined to be

$$\theta = \tfrac{1}{2} \tan^{-1} (u/q) \qquad (2.17)$$

and in this case is equal to ψ. Either pair, p and θ or q and u, can be used to specify the linear polarization; the latter is more fundamental in the sense that q's and u's of successive slabs are additive (eqn (2.11)).

The polarization efficiency in a uniform slab is simply $p/\tau = \Delta\sigma/\delta = (C_{eI} - C_{eII})/(C_{eI} + C_{eII})$.

A combination of two slabs. For two slabs i and j, i being nearer the observer, we have

$$q_{ij} = q_i + q_j \, , \qquad u_{ij} = u_i + u_j \qquad (2.18)$$

and

$$v_{ij} = e_i \, (u_j \cos 2\theta_i - q_j \sin 2\theta_i) \qquad (2.19)$$

where

$$e_i = \int_i \Delta\epsilon \, ds \, . \qquad (2.20)$$

Several interesting phenomena may be noticed even in this elementary model. Because of the disalignment of the two slabs p_{ij} is smaller than the value for $\phi=0$ by a depolarization factor D given by

$$D^2 = (1 + 2r \cos 2\phi + r^2)/(1 + r)^2 \leqslant 1 \qquad (2.21)$$

where $r = p_i/p_j$ and $\phi = \theta_i - \theta_j$. The position angle, given by

$$\tan 2(\theta - \theta_j) = r \sin 2\phi \, / \, (1 + r \cos 2\phi) \, , \qquad (2.22)$$

is wavelength dependent if r is. A rewriting of eqn (2.19),

$$v_{ij} = -e_i \, p_j \, \sin 2\phi \; , \tag{2.23}$$

emphasizes the analogy of a waveplate and also shows how the wavelength
dependence of circular polarization is the product of the effects of birefringence and dichroism. If the grains are similar in the two slabs this
may also be cast in the form

$$v_{ij} = (e_i/p_j) \, p_{ij}^2 \, G \; . \tag{2.24}$$

This is useful because e_i/p_i can be predicted from grain-model calculations
and p_{ij} is an observable, so that ϕ may be determined   from

$$G = -r \sin 2\phi \; / \; (1 + 2r \cos 2\phi + r^2) \; , \tag{2.25}$$

a factor describing the geometry of the alignment in the interstellar medium. There are a number of interrelations such as

$$d\theta/d\lambda = -\tfrac{1}{2}G \; d\ell n \; r/d\lambda \tag{2.26}$$

which may also be useful in interpreting observations.

2.2. CROSS-SECTIONS FOR SINGLE PARTICLES

The basic problem involves the scattering of a plane, linearly polarized,
monochromatic electromagnetic wave by a particle. One must find electric
and magnetic vector solutions to the vector wave equation, or Maxwell's
equations, subject to certain boundary conditions. For the incident wave
there are two separate vector wavefunctions, for each of two orthogonal
polarization modes; each of these modes requires in general two modes in
the interior and scattered waves in order to satisfy the boundary conditions. According to eqns (2.5) and (2.6), calculation of single-particle
cross-sections or efficiency factors, the problem at hand, amounts to finding the far-field component of the forward-scattering amplitude function
which has the same linear polarization as the incident wave.

There are a number of ways in which this can be carried out in practice. The methods applicable to the most general range of particle size
and refractive index are mathematically quite involved and invariably lead
to complex computations requiring a computer. We shall be content to provide a brief description of each approach, avoiding the introduction of
mathematical notation which will serve no further purpose in later chap-

ters. However, there are various approximations which are generally useful
in understanding the behaviour of the exact results, and which find application to certain aspects of interstellar dust. These will be outlined in
§2.3.

2.2.1. Differential equation formulation

This method has been used for particles with boundaries corresponding to a
complete co-ordinate surface in some suitably chosen co-ordinate system.
The solution for scattering by a homogeneous sphere, the Mie theory, is the
most straightforward and the most widely used. Calculations for infinite
circular cylinders are also relatively simple, and have been used extensively to interpret interstellar polarization. Treatment of homogeneous
spheroids, both prolate and oblate, has been developed only recently, and
requires considerably more computational expense.

   The vector solutions are not obtained directly, but are instead derived from solutions of the corresponding scalar wave equation. The
required scalar (or vector) wavefunctions can be expressed as Fourier
expansions in the complete (but not necessarily orthogonal) set of scalar
(or vector) wavefunctions. Boundary conditions are utilized in two ways.
First, the radial wavefunctions are required to exhibit a certain behaviour
at the centre of the particle and at infinity, allowing the alternative
radial wavefunctions with undesirable singular or divergent properties to
be eliminated. Second, at the boundary of the particle the tangential components of electromagnetic field vectors must be continuous, imposing
related continuity conditions on the scalar wavefunctions and their derivatives. This second class of conditions is used to determine the expansion
coefficients of the interior and scattered waves, the coefficients of the
incident wave being known.

   For the sphere, infinite cylinder, and problems with incidence along
an axis of symmetry, orthogonality properties of the angular wavefunctions
can be exploited to yield closed formulae for the expansion coefficients
term by term. For spheroids the orthogonality property has to be obtained
by expanding each spheroidal angular wavefunction and its derivative as an
infinite series of associated Legendre polynomials with known coefficients.
Thus the required expansion coefficients corresponding to the different
orders of the elementary wavefunctions are coupled. Having to solve a set
of simultaneous linear equations and to evaluate so many auxiliary functions escalates the computing time considerably. The vector wavefunctions
of interest are obtained by evaluating the radial wavefunctions for the
scattered wave in the far-field limit. Actually the full angular prescrip-
tion of the scattered wave is found. Although only the forward direction
is used here, the general solution has many other uses, which will be des-

cribed in §4.2.1.

Some of these points can be illustrated by reference to scattering by
a sphere, for the particular case in which the wave is polarized with the
electric vector perpendicular to the scattering plane (component 1 in Fig.
4.2). The relevant scattering amplitude is

$$E_{s1} = -i\{\exp(-ikr)/kr\} E_1 \times$$
$$\sum_{n=1}^{\infty} \{(2n+1)/n(n+1)\}\{a_n \pi_n(\cos\theta) + b_n \tau_n(\cos\theta)\} \qquad (2.27)$$

where

$$\pi_n(\cos\theta) = P_n^1(\cos\theta)/\sin\theta \quad , \quad \tau_n(\cos\theta) = dP_n^1(\cos\theta)/d\theta \quad , \qquad (2.28)$$

a_n and b_n (the Mie coefficients) are the expansion coefficients for the
scalar wavefunctions of the scattered wave, and θ is the scattering angle.
The functions P_n^1 are sufficient for the expansion of the incident wave;
their derivatives arise in the construction of the electric vector from the
scalar wavefunctions. The r-dependent terms come from the asymptotic limit
of the radial wavefunction for the scattered wave, a spherical Hankel function. Comparison of eqn (2.27) with eqn (2.3) gives $S_1(\theta)$. Evaluation of
$S_1(\theta)$ for $\theta=0$ then leads to

$$Q_e + iQ_p = (C_e + iC_p)/\pi a^2 = 2x^{-2} \sum_{n=1}^{\infty} (2n + 1)(a_n + b_n) \qquad (2.29)$$

where $x=2\pi a/\lambda$ is the size parameter for radius a. This result for a sphere
is of course independent of the state of polarization, and so the subscript
1 has been dropped.

In practical evaluations of the cross-sections the infinite series
are convergent and are truncated when the desired accuracy is reached.
Considerable care must be taken to compute the elementary wavefunctions and
the expansion coefficients to high accuracy if reliable cross-sections are
to be obtained, especially for spheroids. Similar techniques have been
applied to obtain solutions for coated or 'core-mantle' particles, such as
the concentric sphere and the coaxial infinite circular cylinder of two
layers. However, a solution for a confocal spheroid has not yet been
worked out. Such models may have some application to interstellar grains,
for example if the particles form in one environment and at a later stage
in their evolution accrete a surface layer of different composition (Chapter 11).

2.2.2. Integral equation formulation
The integral equation formulation has served mainly as a useful theoretical

tool in such roles as demonstrating certain reciprocity relations and deriving the important optical theorem (eqn (2.5)). However, as the basis for
the Fredholm integral method, it has taken on new importance as a powerful
computational technique. A few of the principles and applications will be
mentioned here.

The basic equation is the usual integral equation for the electric
field vector which incorporates an integration with the dyadic Green's
function over the volume of the scattering particle. An auxiliary equation
for the vector scattering amplitude of interest is obtained from the far-
field limit of the basic equation. This amplitude function depends only on
the field within the particle, and so a Fourier representation of the field
is adopted to good advantage. Direct solution of the basic equation is
fundamentally difficult because the kernel is singular. The main thrust of
the new approach is to identify a transformation such that the singularity
can be removed analytically.

The net result is that the transformed basic equation and auxiliary
equation become two coupled Fredholm integral equations. By expressing the
integrals as numerical quadratures the integral equations can be converted
to a system of algebraic linear equations for which the computational solution is straightforward. Evaluation of the matrix elements for this system
is itself quite complex, but many of the calculations suffice for all
directions and polarizations of the incident wave. Although a separate
solution of the linear equations is required for each incident wave speci-
fication, the solution for any one wave does allow computation of the corresponding scattering amplitudes for all directions of scattering. This is
valuable for the applications outlined in Chapter 4. An important feature
of the method is that the derived scattering amplitude satisfies the
Schwinger variational principle, guaranteeing numerical stability.

This method is sufficiently general that particles of many shapes and
particles of inhomogeneous composition can be treated. In practice computations have been restricted to homogeneous particles to simplify evaluation of the matrix elements. The scheme has been tested successfully
against previous results for spheres, spheroids, and infinite circular cylinders. Its versatility has been demonstrated in solutions for ellipsoids
and infinite elliptical cylinders.

To provide some perspective it should be pointed out that the differential equation solutions for spheres (the Mie theory) and infinite circular cylinders are a few orders of magnitude less expensive computationally
than any method for spheroids. Thus for spheres an average cross-section
for a distribution of particle sizes is often calculated; similarly, size
and orientation distributions are often adopted for infinite circular cylinders. The particular shape chosen must of course be governed by the

demands of the application, and so each method has its own particular
value. Yet another approach is described below.

2.2.3. The discrete dipole array

The grain is simulated by an array of N polarizable elements in a vacuum.
A cubic lattice arrangement is chosen so that the polarizability of each
element can be derived from the bulk dielectric constant of the grain
material using the Clausius-Mossotti relation. The outer boundary of the
lattice can approximate any shape; results for spheres, cubes, and a variety of bricks (rectangular parallelipipeds) have been presented. Even the
effects of impurity centres can be taken into account by assigning a reso-
nant polarizability to some of the elements.

Each element is polarized by the combined fields of the specified
incident wave and the scattered waves from the other elements, which radiate as dipoles. Each oscillator is therefore strongly coupled to all the
others and a self-consistent solution must be sought. The resulting system
of $3N/2$ complex simultaneous equations to be solved led to a practical restriction on N of about 100 in early computations by this method. The phase
difference between sites must be small, typically less than 0.5 rad. These
two restrictions effectively limit the size parameter, i.e. the ratio of
grain size to wavelength. Nevertheless the range permitted is large enough
to span the optical spectrum for interstellar grains. Such a lattice
representation is fine enough that Bragg reflections are not a problem as
long as the refractive index of the bulk material exceeds about 1.3. (A
recent reformulation of this method as a variational principle allows N to
be as large as 10^4.)

Once the self-consistent complex amplitudes of the individual dipoles
are determined the summation to find the total far-field forward-scattering
amplitude, and hence the complex cross-section, is straightforward. This
calculation can be carried out for various orientations of the propagation
vector and incident electric vector relative to the grain. The power
absorbed internally by the dipoles can also be summed to give the absorption cross-section. The cross-section for scattering follows by subtraction. In §4.2.2 the treatment of scattering properties using this approximation is mentioned.

2.2.4. Laboratory measurements

Direct measurements in the laboratory serve two purposes: to assess the
validity of the numerical procedures, and to circumvent the problems of
numerical calculations, particularly for complex shapes. Experiments on
extinction and polarization at optical and ultraviolet wavelengths have
been carried out in a small number of cases using actual particles of the

appropriate dimensions. Phase-shift effects are not observed. Some other
limitations of this approach are the difficulties in controlling and measuring the size, shape, and composition of the particles, and the large
amount of experimental time required to produce data covering a sufficient
range of all the parameters involved. Often only normalized curves of the
wavelength dependence of extinction (and rarely polarization) are found,
rather than actual particle cross-sections. Nevertheless this method
serves an important role, particularly in spectroscopic applications.

Results for graphite smoke indicate a rise in far-ultraviolet extinction which is more rapid than the prediction from Mie theory calculations.
The origin of the discrepancy may lie in the adopted bulk refractive index,
but final resolution of this important question is still awaited (§10.2.1).
Other experiments on small metal oxide grains have been used to explore the
behaviour of ultraviolet extinction in the presence of surface excitons.
Measurements on small particles have also contributed mass absorption coefficients near interesting infrared resonances. Comparison of these results
with theoretical calculations also confirms the importance of grain size
and shape and the presence of structural disorder in determining the exact
profile of the resonance.

Another technique involves setting up a microwave analogue of the
interstellar scattering process in the laboratory. Measurements are made
at centimetre wavelengths on centimetre-sized model grains. The material
for the models is chosen to have a centimetre-wavelength complex refractive
index in the range expected for interstellar grains at optical wavelengths.
Different size parameters are obtained by changing the model size rather
than the wavelength of the radiation. The complex cross-section is determined from the measured forward-scattering amplitude. By movement of the
microwave detector the differential scattering cross-section can be measured. In principle the total scattering cross-section could be obtained
by integration over all solid angles and then by subtraction the absorption
cross-section would follow.

The actual measurement techniques are quite elaborate and so the
method is not widely used. Observations of smooth spheres have been made
for calibration and for estimating the experimental accuracy. The main
contributions of interest here have been to the understanding of the forward scattering by roughened spheres and by elongated particles such as
spheroids and finite circular cylinders.

2.3. APPROXIMATION FORMULAE

For certain limited ranges of size, wavelength, and refractive index there
are relatively simple analytical formulae for the complex extinction

cross-section. Some of these have applications in the study of interstellar dust. Insight into the results of the exact calculations is also provided.

2.3.1. The long-wavelength limit

By this phrase we mean that the wavelength of radiation exceeds all dimensions of the particle ($x \ll 1$), and in addition that

$$|mx| \ll 1 . \tag{2.30}$$

The relevant formulae are therefore most useful for the infrared spectral region. There are two methods for finding the approximate solutions. The first considers the particle as an electric dipole oscillating, and therefore radiating, in a spatially homogeneous applied field with components E_{0j} ($j=1,2,3$). The dipole moment induced is

$$P_i = \alpha_{ij} E_{0j} \tag{2.31}$$

where α_{ij} is the polarizability tensor for the particle ($i=1,2,3$). To calculate the complex cross-sections using eqns (2.5) and (2.6) we need to know only the amplitude of that component of the forward-scattered dipole radiation that has the same polarization as the incident wave. Suppose the incident wave is travelling along an axis labelled 3 and is polarized with the electric vector along direction j ($j=1,2$); then the relevant component arises from α_{jj} and

$$C_j = C_{ej} + iC_{pj} = 4\pi k i \, \alpha_{jj}. \tag{2.32}$$

Consider how these values of α_{jj} are related to the two principal components of polarizability of an axially symmetric particle. Let α_A and α_B be the components for the two equal axes and for the symmetry axis respectively. Without loss of generality we may choose the symmetry axis to lie in the 2-3 plane, inclined at angle ζ with respect to axis 3. (Note that axes 1 and 2 are equivalent to the directions I and II of the grain profile in Fig. 2.1 only for oblate particles; for prolate particles axis 2 lies along I. We shall adhere to this 'symmetrical particle convention' for the axes throughout, in particular in §§ 2.4.2, 2.4.3, and 2.4.4.) Then

$$\alpha_{11} = \alpha_A \quad \text{and} \quad \alpha_{22} = \alpha_B \sin^2\zeta + \alpha_A \cos^2\zeta \tag{2.33}$$

so that in the notation of eqn (2.32)

$$C_1 = C_A \quad \text{and} \quad C_2 = C_B \sin^2\zeta + C_A \cos^2\zeta \; . \tag{2.34}$$

This dependence on inclination ζ is not valid outside the limits of the long-wavelength approximation.

For any ellipsoid of volume V the principal components α_Z (Z=A,B,C) can be calculated from the depolarization factors L_Z using

$$\alpha_Z = (V/4\pi)[(m^2-1)/\{L_Z(m^2-1) + 1\}] \; ; \tag{2.35}$$

some formulae for L_Z for spheroids are given in Table 2.1. Small cylinders and disks are limiting cases. The corresponding efficiency factors Q_Z are given as well. Note that the factors vary as λ^{-1} for constant m, as will be seen below in the discussion of the computed curves. However the extinction (the real part) vanishes in this approximation if m is real because only absorption, as distinguished from scattering, has been accounted for. The remaining entries in Table 2.1 are discussed in §4.4.1, where the λ^{-4} dependence of the scattering cross-section for small particles is explained.

Infinite cylinders do not properly belong in the above discussion because the wavelength does not exceed their length (condition (2.30)). However, there is a second approach by which approximate solutions may be obtained. The formulae for the expansion coefficients (e.g. the Mie coefficients) in the general solution for the scattering amplitudes are expanded as power series in the size parameter x. Then the cross-sections are evaluated using equations such as eqn (2.29) for C. Normally it is only useful to keep the lowest-order term in each series; the range of validity of series with a few more terms is quite limited and then the exact expressions must be used. Such a procedure has been carried out for spheres and for infinite circular cylinders.

For spheres it is not surprising that the first term in C gives the result already found above. Higher-order multipole moments are included only when more terms are retained. For infinite cylinders, it is found that the expressions for C are actually the same as those tabulated for thin short cylinders and that eqn (2.34) applies.

2.3.2. Anomalous diffraction
Inspection of plots of the frequency dependence of the efficiency factors of particles with different values of m reveals some common features. This becomes particularly clear when the parameter $\rho = 2x(n-1)$ is used as abscissa, as shown in the figures in §2.4. The particular limit in which the size parameter x is increased and $|m-1|$ is decreased while ρ is held fixed is called anomalous diffraction. In this regime simple solutions for

TABLE 2.1

APPROXIMATE EXPRESSIONS IN THE LONG-WAVELENGTH LIMIT

Shape	L_B	L_A	Normalization Area	Volume	Q_A §	Q_B	Q_{SA} ‡	Q_{SB}
Sphere (a=b)	$\frac{1}{3}$	$\frac{1}{3}$	πa^2	$\frac{4}{3}\pi a^3$	$i4x\left(\frac{m^2-1}{m^2+2}\right)$	$i4x\left(\frac{m^2-1}{m^2+2}\right)$	$\frac{8}{3}x^4\left|\frac{m^2-1}{m^2+2}\right|^2$	$\frac{8}{3}x^4\left|\frac{m^2-1}{m^2+2}\right|^2$
Oblate spheroid (a>b)	$\frac{1+f^2}{f^2}\left[1-\frac{1}{f}\tan^{-1}f\right]$, $f^2=a^2/b^2-1$	$\frac{1-L_B}{2}$	πab	$\frac{4}{3}\pi a^2 b$	$\frac{i4x(m^2-1)}{3L_{A,B}(m^2-1)+3}$	$\frac{i4x(m^2-1)}{3L_{A,B}(m^2-1)+3}$	$\frac{8}{3}x^3y\left|\frac{m^2-1}{3L_{A,B}(m^2-1)+3}\right|^2$	same form as oblate
Thin disk (a≫b)	1	0	$2ab$	$\pi a^2 b$	$\frac{i\pi}{2}x(m^2-1)$	$\frac{i\pi x(m^2-1)}{2m^2}$	$\frac{1}{12}x^3y\,|m^2-1|^2$	$\frac{1}{12}x^3y\left|\frac{m^2-1}{m^2}\right|^2$
Prolate spheroid (a<b)	$\frac{1-e^2}{e^2}\left[-1+\frac{1}{2e}\ln\frac{1+e}{1-e}\right]$, $e^2=1-a^2/b^2$	$\frac{1-L_B}{2}$	πab	$\frac{4}{3}\pi a^2 b$	same form as oblate			
Thin cylinder (a≪b)	0	$\frac{1}{2}$	$2ab$	$\pi a^2 b$	$\frac{i\pi}{2}x\frac{2(m^2-1)}{m^2+1}$	$\frac{i\pi x(m^2-1)}{2}$	$\frac{1}{12}x^3y\left|\frac{2(m^2-1)}{m^2+1}\right|^2$	$\frac{1}{12}x^3y\,|m^2-1|^2$
Infinite cylinder	$-$	$-$	$2a$ ‡	πa^2	same form as thin cylinder		$\frac{\pi^2 x^3}{16}\left|\frac{2(m^2-1)}{m^2+1}\right|^2$	$\frac{\pi^2 x^3}{8}\left|m^2-1\right|^2$

§ $x=2\pi a/\lambda$ ‡ $y=2\pi b/\lambda$ ‡ Cross-sections per unit length are computed.

the cross-sections may be derived. They are computationally convenient to
use and adequately describe many of the features observed for normal parti-
cles with the same ρ but finite $|m-1|$. On the other hand, certain informa-
tion has to be sacrificed for this ease of solution. For example the angular distribution of the scattered (optical) radiation is not at all well
approximated because x is so large compared with the value for a normal
particle.

Since x is large the solution can be based on the ray approximation.
Also in this regime $|m-1| \ll 1$, so that the ray may be traced straight
through the particle with no refraction or reflection. Unfortunately with
these simplifications any dependence of the cross-sections of the normal
particle on the polarization of the incident wave is lost. (For example
the trends $\Delta\sigma \to 0$ and $\Delta\epsilon \to 0$ as $m \to 1$ will be noticed in the figures below.)
This incidentally points out an alternative approach to the approximate
solution; it could be derived directly from the scalar (rather than the
vector) wave equation and hence the name scalar wave approximation.

The method of solution depends on specifying the transmission along
rays and then treating the diffraction according to Huygens' principle.
The important factors are the amounts by which the amplitude and intensity
change with pathlength p along the ray. These may be written $c(p) =$
$\exp(-i\rho^{\dagger}p/2a)$ and $d(p) = \exp(-\gamma p)$, respectively, where $\rho^{\dagger} = 2x(m-1)$, a is
the semi-axis for which the size parameter is defined, and $\gamma = -\mathrm{Im}(\rho^{\dagger})/a$ is
the linear absorption coefficient. The forward-scattered amplitude (and
hence the complex cross-section) is proportional to the integral of 1-c
over the geometrical shadow area. The cross-section for absorption is proportional to the integral of 1-d. Once C_e and C_a (for absorption) are determined, C_s (for scattering) follows by subtraction.

Evaluation of the integrals may proceed by numerical integration or
by substitution in closed-form solutions for particularly simple shapes.
For example for the sphere

$$Q = Q_e + iQ_p = 4\,K(i\rho^{\dagger}) \quad \text{and} \quad Q_a = 2\,K(\gamma 2a) \qquad (2.36)$$

where the common integral form is

$$K(w) = \int_0^{\pi/2} \{1-\exp(-w\sin\tau)\}\cos\tau\;\sin\tau\;d\tau$$

$$= \tfrac{1}{2} + \exp(-w)/w + \{\exp(-w)-1\}/w^2\;. \qquad (2.37)$$

The asymptotic limits for large ρ are Q=2 and, if $\gamma \neq 0$, $Q_a = 1$. Note that the
latter limit does not hold generally, since if $|m-1|$ is at all significant
there is reflection (§4.4.3). For small ρ series expansion gives

$$Q_e = 4\gamma a/3 + \{\rho^2 - (\gamma a)^2\}/2 \quad \text{and} \quad Q_a = 4\gamma a/3 - (\gamma a)^2 \; ; \qquad (2.38)$$

by subtraction $Q_s = |\rho^+|^2/2$. Notice the ways in which the wavelength dependence differs from that in the long-wavelength approximation.

The formulae for the efficiency factors for a spheroid are the same as those for a sphere if the radius a is replaced by half the length of the central ray through the spheroid. The normalization area for Q is the geometrical shadow area. Because the exact solutions for spheroids have become available only recently, these approximations have played an important role. In the case of infinite cylinders closed-form expressions involving ordinary and modified Struve and Bessel functions have been found; however, in practical applications direct numerical integration is an efficient alternative. Asymptotic formulae for small ρ are quite similar to eqn (2.38).

The anomalous diffraction approximation for Q_e (eqn (2.36)) is plotted in Figs. 2.2 and 2.3 for comparision with Mie theory results for finite $|m-1|$. The principle feature of the curves at intermediate values of ρ is a series of maxima and minima as ρ increases. In the anomalous diffraction theory this is traced to constructive and destructive interference between transmitted and diffracted radiation. (Extinction is caused by interference of both of these components with the incident wave.) The phase shift depends on the transmitted wave because the diffracted wave is always in anti-phase with the incident wave; therefore Q_p is zero near both the maxima and minima in Q_e (Fig. 2.4). As absorption within the particle is increased, the transmitted wave is attenuated and the interference effects in Q_e are seen to disappear.

It is also clear that for intermediate values of ρ the anomalous diffraction approximation is a fair representation of the behaviour for finite $|m-1|$. In Fig. 2.4 anomalous diffraction results are included to make the same point for spheroids (the same could be done for infinite cylinders). Two trends will be noticed in the figures: the agreement improves as $|m-1|$ decreases, and for the non-spherical particles the approximation is best for that orientation in which the central ray is the shortest. This emphasizes that any refinement of this approximation would have to deal with grazing reflection, total internal reflection, refractive bending, and surface waves.

There is at least one practical application: the study of X-ray scattering and absorption by interstellar grains. Usually the conditions $x \gg 1$, $|m-1| \ll 1$, and $|\rho^+| \ll 1$ hold, so that asymptotic formulae like eqn (2.38) may be applied. Furthermore, since the condition $x \gg 1$ is a correct physical description rather than an artefact of the approximation theory, the angular distribution of the scattered radiation is correctly described as a small-angle diffraction-like phenomenon (§4.4.2).

2.3.3. Rayleigh-Gans scattering

The formulae required for X-ray scattering can also be found from another
approximation regime, called Rayleigh-Gans scattering, for which the mini-
mum conditions $|m-1| \ll 1$ and $|\rho^{\dagger}| \ll 1$ must hold. The basic approach, for
a particle with either small or large x, is to find the scattering amplitude produced by each volume element independently using the long-wavelength (Rayleigh) approximation. This can be done simply because the volume
elements are uncoupled by the condition $|\rho^{\dagger}| \ll 1$; the incident wave seen
by each volume element is the same, and the scattered wave from each
escapes unaffected also. (Compare this with the strong coupling encountered in the discrete dipole array method.) The total scattering amplitude
is obtained by integrating the individual scattering amplitudes over the
particle volume, taking proper account of phase differences which produce
interference. Determining the contribution from absorption is a straightforward summation over independent volume elements. Note that the efficiency factors found are necessarily always very small.

For application to X-ray scattering these formulae can be reduced
further using the additional condition x >> 1; the results are of course
identical to those from the asymptotic limit of the anomalous diffraction
approximation.

2.4. SOME CHARACTERISTICS OF COMPLEX CROSS-SECTIONS

The results of some computations of complex cross-sections for single particles are presented here to illustrate some of the important characteristics and their relevance to the interpretation of interstellar extinction
and polarization. The particles have sizes comparable with the wavelength
so that calculations had to be performed by one of the general methods described in §2.2; the various differential equation formulations were actually used. For simplicity it has been assumed that the complex refractive
index of the material, $m=n-in'$, is a constant. This is not necessarily a
good assumption; even for the many materials such as water ice, silicates,
and graphite for which m is reasonably constant across the optical spectrum, significant changes do occur in the infrared and ultraviolet. Therefore the curves displayed should not be applied indiscriminately.

2.4.1. Spheres and extinction

Curves of Q_e for non-absorbing (dielectric) spheres ($n'=0$) with different n
are plotted in Fig. 2.2. Use of ρ (which is proportional to λ^{-1} with our
assumption) as abscissa brings out the valuable scaling rules for a, λ, and

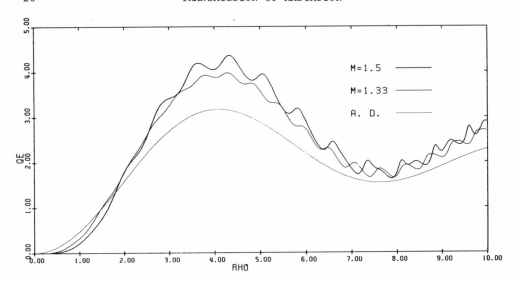

FIG. 2.2. The ρ dependence of the efficiency factor for
extinction by a dielectric sphere for two refractive indices.
The anomalous diffraction (A.D.) approximation is also
included.

n-1 embodied in ρ. The similarity with the anomalous diffraction approximation is also clear. Over a substantial range before its first maximum at
ρ≈4, Q_e varies as ρ. A similar λ^{-1} portion is seen in the interstellar
extinction curve at optical wavelengths. Recognition of this correspondence is what allows an estimate of the grain size; the important general
conclusion is that interstellar dust particles are very small, their characteristic dimension being of the order of 0.1 μ. (For example ρ(V) ≈ 2
(incidentally Q_e(V) ≈ 2), so that a(n-1) ≈ 0.09 μ; and a ≈ 0.25 μ if n ≈
1.33.)

      In Fig. 2.3 the effects of absorption (n′>0) are shown. The main
effect on the λ^{-1} portion is to shift it to smaller ρ, so that for a given
n-1 the required particle size decreases as n′ increases.

      Interference effects similar to those discussed for anomalous dif-
fraction produce a series of broad maxima in the upper curves beyond ρ≈4.
It can be seen that when the transmitted radiation is suppressed by absorption in the particle (n′>0) these oscillations are damped. The superimposed ripples are also interference phenomena, dependent on n-1, and they
too disappear when n′>0. It should be noted that in interstellar extinction (and polarization) one should not expect to see these interference
phenomena, even if n′≈0, since the interstellar particles no doubt will
depart from perfect symmetry and will vary in size (and will have different
orientations relative to the observer). If desirable, these distracting

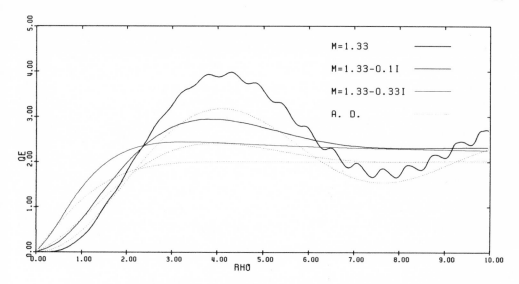

FIG. 2.3. Changes in the efficiency factor for extinction by a
sphere as the imaginary part of the refractive index is
increased (see also Fig. 4.3). The appropriate anomalous dif-
faction (A.D.) approximation is shown for each case.

effects can be smoothed out in computed curves by averaging over a size (or
orientation) distribution.

For small ρ the behaviour of Q_e predicted by the long-wavelength
approximation can be seen, i.e. a λ^{-4} dependence for $n'=0$ and a λ^{-1} dependence for $n'>0$. (Note that asymptotic behaviour of the anomalous diffrac-
tion approximation is different for $n'=0$.) Since the λ^{-1} portion has been
identified with the optical spectrum, typical applications of the long-wavelength approximation may be expected to be in the infrared.

At large ρ (small λ) all of the curves approach 2, so that the
cross-section is actually twice the projected area. Reflected, refracted
and absorbed components account for a value unity. The doubling is
explained by diffracted radiation which, although strongly peaked in the
forward direction, is in effect removed from the propagating plane wave.
The characteristic flattening beyond the λ^{-1} portion might be expected to
appear in ultraviolet observations of extinction, but on the contrary a
continued rise in extinction has been found. In retrospect it is not difficult to contemplate the addition of particles for which the first maximum
in Q_e is in the ultraviolet; these particles would contribute little to
extinction at optical wavelengths. Taking into account how the theoretical
extinction curves scale with ρ it is tempting to adopt the terminology
'small' for these additional particles, to distinguish them from the
'normal' grains causing optical extinction. For convenience this crude

separation will be accepted at present; later the physical basis for
expecting a wide range in particle parameters (this would include a simple
bimodal distribution) will be explored.

2.4.2. Spheroids, infinite cylinders, and polarization
Calculations for prolate and oblate spheroids are useful for interpreting
interstellar polarization. (The polarizing particles also produce extinction, perhaps all of it. We shall not address this directly, except to
note that the features in the extinction curve are similar to those for
spheres, as can be seen by inspecting the figures.) Because the solution
for infinite circular cylinder cross-sections has been available so much
longer, and is so much less time consuming (expensive) to compute, much of
the interpretation of interstellar polarization has made use of infinite
cylinder, rather than spheroid, results. As will become clear, this
approximation is quite suitable for many semi-quantitative considerations.

     First, we give a brief description of the terminology. The sub-
scripts 1 and 2 used here for the two polarization modes follow the convention for symmetrical particles set up in §2.3.1, and correspond respectively to the following alternatives used in the literature: TE and TM
(transverse electric and transverse magnetic); H and E; case II and case I,
the latter two being peculiar to infinite circular cylinders. Recall that
ζ measures the angle between the symmetry axis and the direction of propagation, and that a is the radius for the equal axis and 2b is the dimension
along the symmetry axis. The abscissa in all diagrams is the ρ corresponding to dimension a. To obtain the efficiency factors, cross-sections for
spheroids are normalized by πab and for infinite circular cylinders by $2a$.
These choices of geometrical factors and ρ stem from the important case of
perpendicular or edge-on incidence ($\zeta=90°$).

     Interstellar particles are not statically aligned (picket-fence
alignment), but rather are thought to be tumbling about a preferred direction in space which projects into axis II in the plane of the sky (dynamic
alignment). Therefore the cross-sections for grains with various orientations should be averaged. However, many of the salient features can be
demonstrated using single particles.

     We begin by examining some typical results for oblate and prolate
spheroids, displayed in Fig. 2.4. With the exception of extra curves for
Q_e and Q_p for the case $\zeta=0°$ all the efficiency factors apply to $\zeta=90°$. Our
first comment concerns the anomalous diffraction approximation which is
also shown for both Q_e and Q_p with both $\zeta=0°$ and $90°$. When the direction
of incidence is along the short direction of the particle the approximation
is not bad (cf. Fig. 2.2); however, along the long direction it is unac-
ceptable, even with n as low as 1.33. Therefore anomalous diffraction

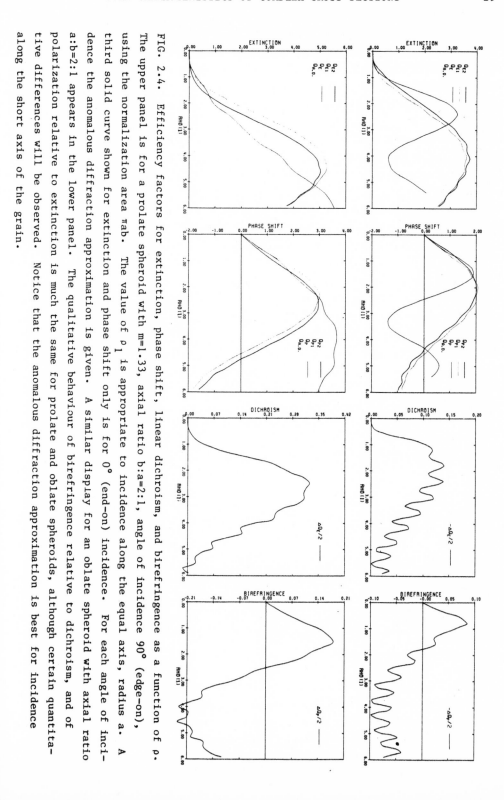

FIG. 2.4. Efficiency factors for extinction, phase shift, linear dichroism, and birefringence as a function of ρ. The upper panel is for a prolate spheroid with m=1.33, axial ratio b:a=2:1, angle of incidence 90° (edge-on), using the normalization area πab. The value of ρ₁ is appropriate to incidence along the equal axis, radius a. A third solid curve shown for extinction and phase shift only is for 0° (end-on) incidence. For each angle of incidence the anomalous diffraction approximation is given. A similar display for an oblate spheroid with axial ratio a:b=2:1 appears in the lower panel. The qualitative behaviour of birefringence relative to dichroism, and of polarization relative to extinction is much the same for prolate and oblate spheroids, although certain quantitative differences will be observed. Notice that the anomalous diffraction approximation is best for incidence along the short axis of the grain.

results are not recommended for studying the orientation dependence of
extinction by dielectric particles.

Next we consider the polarization, noting that $\Delta\sigma$ and $\Delta\epsilon$ are proportional to the plotted quantities $\frac{1}{2}\Delta Q_e$ and $\frac{1}{2}\Delta Q_p$. Notice the maximum in ΔQ_e
at $\rho \simeq 2$, and recall that on the basis of the extinction curve $\rho \simeq 2$ falls
within the optical region of the spectrum. Therefore we might expect to
observe a maximum in interstellar linear polarization at optical wavelengths, which is indeed the case. The wavelength at which this maximum
occurs is called λ_{max} (hence also ρ_{max}).

At about λ_{max} the birefringence $\frac{1}{2}\Delta Q_p$, and thus any optical circular
polarization produced, changes sign. Observationally it is found that the
cross-over wavelength λ_c is near λ_{max}, just as $\rho_c \simeq \rho_{max}$ in these particular results. We shall now see how the placement of ρ_{max} and the ratio

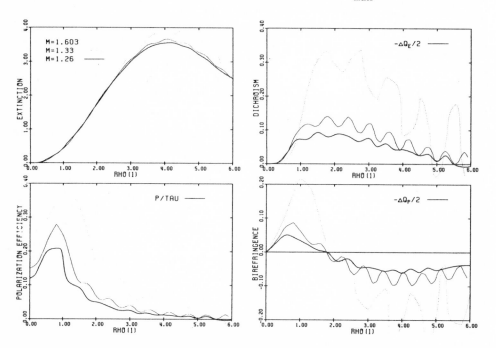

FIG. 2.5. Efficiency factors and linear polarization efficiency for dielectric prolate spheroids with axial ratio b:a=2:1
and angle of incidence 90°. The extinction efficiency factors
are the average for natural light, and show the dependence on
refractive index revealed in Fig. 2.2. As the refractive
index increases so does the amount of polarization. However,
important characteristics like ρ_{max}, ρ_c, and the width of the
dichroism peak do not change significantly. (Curve fitting for
the case m=1.603 suffers from undersampling.)

λ_c/λ_{max} depend on other possible choices of the parameters of the spheroids. Also the quantity p/τ, which measures the polarizing efficiency (§2.1.3), will be plotted (instead of the values of Q_p which cannot be measured anyway).

The anomalous diffraction results, which deal with the limit $n \to 1$, give no polarization. Conversely, it is clear from the results in Fig. 2.5 that the polarizing efficiency increases with $n-1$. Thus for a given shape and degree of alignment some materials will be more favourable than others. Note also the interdependence of a, λ (and λ_{max}), and $n-1$ through ρ.

For a given material the polarization increases as the particle shape

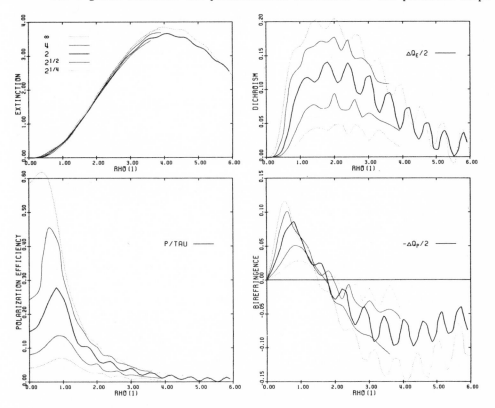

FIG. 2.6. Same as Fig. 2.5, except that the dependence on axial ratio b:a for a single refractive index m=1.33 is shown. Results for an infinite circular cylinder are included, with the ρ scale stretched by a factor 1.13 to align the peaks in extinction (cf. Fig. 2.8); this factor compensates for the different shape, not the axial ratio. The polarization efficiency increases with axial ratio, but is relatively good near ρ_{max} even for a ratio as low as 2:1.

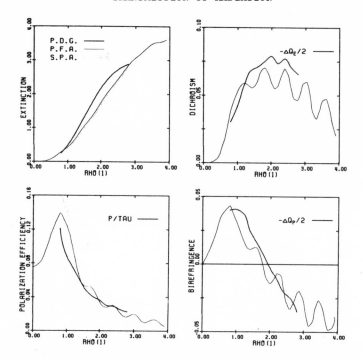

FIG. 2.7. Thick curves show the average efficiency factors and corresponding polarization efficiency for a prolate spheroid spinning around a short axis aligned perpendicular to the direction of incidence (perfect spinning or perfect Davis-Greenstein alignment (P.D.G.)). As in several earlier figures, $m=1.33$ and $b:a=2:1$. A small-particle approximation (S.P.A.) is obtained by using eqn (2.34) averaged for perfect spinning alignment and substituting as the two cross-sections those computed for edge-on incidence. The results for birefringence and dichroism are identical to the picket-fence approximation (P.F.A.) shown, and amount to dividing the edge-on results by 2. Thus it is seen that the polarization reduction factor for perfect spinning alignment relative to picket-fence alignment is about 0.5. Both ρ_{max} and ρ_c increase slightly, but by similar amounts. For extinction the curve for edge-on incidence is plotted, whereas for polarization efficiency eqn (3.15) was used with $|q_{SB}|=0.5$, beyond the actual range of its validity.

becomes more elongated, as shown in Fig. 2.6. (The same applies to flattening of oblate particles.) The results for an infinite circular cylinder are included to help answer the question of the suitability of the infinite cylinder approximation. Apparently the behaviour of the cross-sections is qualitatively quite similar to that of prolate spheroids, even those with

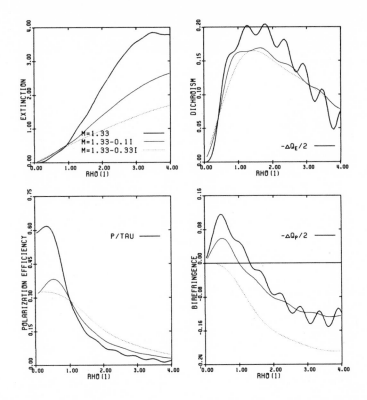

FIG. 2.8. Efficiency factors and polarization efficiency for
infinite circular cylinders at angle of incidence 90°. The ρ
scale is appropriate to the dielectric case, m=1.33. For
m=1.33-0.1i and m=1.33-0.33i the ρ scales have been stretched
by factors of 1.5 and 3.3 respectively to make the peaks in
linear dichroism (ρ_{max}) coincide. This demonstrates that
grains producing much the same wavelength dependence of polarization (and extinction, but see Fig. 2.3) will have characteristically different birefringence, with $\rho_c < \rho_{max}$, if the absorption is significant.

small axial ratios. The polarizing efficiency of a spheroid in the optical
region is relatively good at a ratio of 2:1, and approaches the limit by
4:1. We tentatively conclude that infinite cylinder calculations will be
useful for all but the most detailed quantitative applications.

To this point only edge-on incidence (ζ=90°) has been mentioned.
Fig. 2.7 considers oblique incidence, in particular an average over ζ
called perfect spinning alignment. Although the individual results are not
displayed, it is clear that while the polarization decreases as ζ decreases
from 90° the important qualitative characteristics of the edge-on case,

such as the location of ρ_{max} relative to ρ_c, are preserved. The polarization reduction factor for the average shown is about 0.5, as can be derived
from simple approximations. Generally no fundamental changes for tumbling
particles are expected either, though of course quantitative deductions
will differ somewhat. Oblate particles also show no startling differences,
but there is some variation in the shape of the polarization curves and in
the scaling with ρ.

To end our brief survey we consider how the polarization depends on
n', the imaginary part of the refractive index. Introduction of absorption
affects all of the curves, in the sense of shifting the rise in extinction
to smaller ρ (Fig. 2.8, cf. Fig. 2.3). To the extent that polarization and
extinction shift together, this produces no significant observational
consequences. However, there is one important differential shift, that in
ρ_c relative to ρ_{max}. As n' increases, the whole birefringence curve is
lowered so that $\rho_c/\rho_{max} < 1$. This permits a fundamental distinction between absorbing materials $(n'>0)$ and dielectric materials $(n'/(n-1) \ll 1)$
that can be exploited using measurements of interstellar birefringence
(§3.3).

Our few examples certainly do not exhaust the wide range of possibilities to be explored. They should give an impression of the principal
trends and thereby serve as a background for our discussions of the observations. Among the many refinements that can be introduced in the calculations for comparison with the data is the wavelength dependence of m, which
is quite marked for some materials. This presupposes some knowledge of
what grain materials are present and assumes that laboratory measurements
of m at the low temperatures characteristic of the interstellar medium have
been carried out over a large wavelength range.

FURTHER READING

§2.1

MARTIN, P.G. (1974). Interstellar polarization from a medium with changing
    grain alignment. Astrophys. J. 187, 461.

§2.2.1

VAN DE HULST, H.C. (1957). Light scattering by small particles. Wiley,
    New York.
KERKER, M. (1969). The scattering of light. Academic Press, New York.
WICKRAMASINGHE, N.C. (1973). Light scattering functions for small parti-
    cles. Wiley, New York.
SHAH, G.A. (1970). Scattering of plane electromagnetic waves by infinite
    concentric circular cylinders at oblique incidence. Mon. Not. R.

astron. Soc. 148, 93.

ASANO, S., and YAMAMOTO, G. (1975). Light scattering by a spheroidal par-
    ticle. App. Opt. 14, 29.

§2.2.2

HOLT, A.R., UZUNOGLU, N.K., and EVANS, B.G. (1978). An integral equation
    solution to the scattering of electromagnetic radiation by dielectric
    spheroids and ellipsoids. IEEE Trans. Antennas Propag. 26.

§2.2.3

PURCELL, E.M., and SHAPIRO, P.R. (1977). A model for the optical behaviour
    of grains with resonant impurities. Astrophys. J. 214, 92.

YUNG, Y.L. (1978). A variational principle for scattering of light by die-
    lectric particles. Preprint.

§2.2.4

See §1.2.4 (Huffman 1977).

ZERULL, R., and GIESE, R.H. (1974). Microwave analogue studies. In Plan-
    ets, stars and nebulae studied with photopolarimetry (ed. T. Geh-
    rels). University of Arizona Press, Tucson.

GIESE, R.H., WEISS, K., ZERULL, R.H., and ONO, T. (1978). Large fluffy
    particles: A possible explanation of the optical properties of
    interplanetary dust. Astron. Astrophys. 65, 265.

§2.3.1

See §2.2.1 and

KERKER, M., COOKE, D.D., and CARLIN, J.M. (1970). Light scattering from
    infinite cylinders. The dielectric-needle limit. J. opt. Soc. Am.
    60, 1236.

§2.3.2

CROSS, D.A., and LATIMER, P. (1970). General solutions for the extinction
    and absorption efficiencies of arbitrarily oriented cylinders by ano-
    malous-diffraction methods. J. opt. Soc. Am. 60, 904.

§2.3.3. See §2.2.1 (van de Hulst 1957).

§2.4. See §2.2.1.

3

INTERSTELLAR EXTINCTION AND POLARIZATION

3.1. EXTINCTION

3.1.1. Observations of extinction

The wavelength dependence and amount of extinction are determined by comparing the spectral energy distribution of a reddened star with that of a similar, but unobscured, star. For example, at optical and near-infrared wavelengths intrinsic colours have been extensively studied as a function of spectral type and luminosity class. Therefore for an obscured star whose spectral type and luminosity class is known, measurement of the colours yields colour excesses and hence the extinction curve. An example is shown in Fig. 3.1. The object observed is VI Cyg #12, a highly obscured B8 Ia star with $E_{B-V} = 3.25$. The relative units on the right-hand scale correspond to the measured colour excesses. If measurements are extended far enough into the infrared then the absolute amount of extinction (the zero point) is found because $A(\lambda) \rightarrow 0$ at long wavelengths.

A normalized form of the derived extinction curve is often used for two reasons: first, the deduced colour excesses in the infrared are sometimes uncertain because of imprecise knowledge of intrinsic colours and/or unrecognized contamination by circumstellar emission; second, because the amount of extinction varies from star to star, a normalized curve is needed for intercomparison. Theoretical curves of $Q_e(\lambda)$ may be similarly normalized. Often $E(\lambda-V)/E(B-V)$ is the chosen form, in which case extrapolation

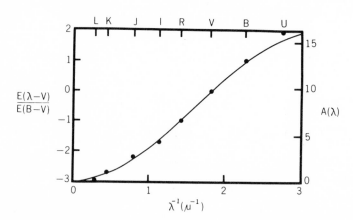

FIG. 3.1. The extinction curve for VI Cyg #12, with the usual normalized ordinate given on the left-hand side and the total extinction in magnitudes on the right-hand side.

of the observed curve to long wavelengths where $E(\lambda-V) = -A_V$ gives as
intercept $-R$ (see the left-hand scale in Fig. 3.1). Conversely, an average
R determined by independent means may be applied to transform a normalized
curve (lacking infrared data) into one on an absolute scale.

The characteristic λ^{-1} behaviour of the extinction in the optical is
indicative of grains with sizes comparable with optical wavelengths
(§3.1.2). Much smaller (or larger) grains would produce too steep (or too
flat) a dependence. Not shown in this figure is a slight enhancement in
extinction at 10 μ ($\Delta A \simeq 0.6$). This and other important spectral features
are discussed in §10.2.

Extinction curves have been measured for stars of a wide variety of
spectral types in all directions in the Galaxy. Data at higher resolution
are also available (§10.2). While there are detectable differences from
star to star, the overwhelming impression is one of uniformity in the
extinction curves. This in turn must reflect considerable uniformity in
grain size and composition, at least within a few kiloparsecs of the sun.
Although the value of R has been a matter of some controversy, a number of
independent techniques seem to point consistently to 3.3 ± 0.3 in the solar
neighbourhood. Determination of R from colour differences, as described
above, has become more reliable with improvements in infrared photometry
and recognition of intrinsic infrared excesses in some types of star
(§7.2.1). The variable-extinction method makes use of the variations in
reddening from star to star in a cluster. If foreground or background
stars are inadvertently treated as cluster members R will usually be overestimated, and so considerable care must be exercised. Two other techniques rely on obtaining the distance to the object independently of R, so
that A_V can be found from the apparent distance modulus. In the cluster-
diameter technique the distance to the cluster is judged from the cluster's
angular size and an assumed linear size, whereas in the kinematical method
the distance to a star is obtained from its radial velocity in the context
of differential galactic rotation. The latter two approaches rely on averaging over a large sample and so possible regional variations are difficult
to detect. The anomalies revealed by the other methods are important
because they offer the possibility of determining what localized conditions
are responsible for the modifications in the grain properties.

Observations of spectral energy distributions in the ultraviolet have
been limited by instrumental sensitivity to bright OB stars in relatively
nearby associations. Differential extinction curves can be derived from
the ratio of the flux of the reddened star to the flux of a comparison
star, the latter being chosen to have approximately the same spectral type
and low E_{B-V}. In practice the spectrum of the comparison star is corrected
to zero reddening using the average extinction curve. Normalized curves of

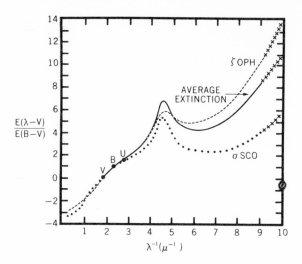

FIG. 3.2.   Normalized extinction curves extended into the ultraviolet. Note the feature at 2200 Å and the continued rise in extinction towards shorter wavelengths. Ultraviolet data from OAO-2 and Copernicus (crosses).

$E(\lambda-V)/E(B-V)$, from OAO-2 and extended by Copernicus, are shown in Fig. 3.2. Extensive satellite observations of colour excesses have also been made.

Substantial variations in shape from star to star, especially in the far ultraviolet, are immediately evident. There are, however, a number of obvious features the curves have in common. The extinction continues to rise through the ultraviolet out as far as it has been measured ($10 \mu^{-1}$); it does not flatten out as might have been expected from the optical data. Therefore a grain model more complex than specifying a single size and composition is indicated. The curves are characterized further by a spectral peak in extinction at $\lambda_o^{-1} = 4.6 \pm 0.1 \mu^{-1}$, and a relative minimum in the range $5.5 - 7.5 \mu^{-1}$. The colour excess $E(\lambda_o -3320)$, which measures the strength of the peak, is generally well correlated with $E(B-V)$, implying that the source of this feature is well mixed with the grains causing optical extinction throughout the interstellar medium.

3.1.2. Interpretation of the extinction curve
The λ^{-1} dependence of extinction at optical wavelengths indicates a predominant contribution from grains with sizes comparable with the wavelength. This can be put on a quantitative basis by finding the parameters of the adopted grain model which best reproduces the extinction data. Unfortunately such an analysis is rather complex and non-unique, since such things as the distributions of grain sizes and shapes, the mixture of grain compo-

sitions, and the character of core-mantle particles would have to be specified. These possibilities will be taken up in Chapters 10 and 11, but at
this stage a very simple-minded approach will suffice. Suppose that there
is only one type of grain with a single size and composition, that the
refractive index is real and constant at optical wavelengths, and that the
theoretical extinction curve can be approximated by the anomalous-diffrac-
tion result for spheres and spheroids (eqn (2.36)). A fit to the data,
shown in Fig. 3.1, implies the correspondence of $\lambda_V^{-1}=1.8 \mu^{-1}$ and $\rho_V=2.2$.
Consequently $a \simeq 0.3 \mu$ if $m \simeq 1.33$, as would be appropriate for 'ice'
grains; silicate grains, with $m \simeq 1.6$, would be about half as large. (More
detailed models give similar results.)

It is clear that to explain the continued rise in extinction into the
ultraviolet more than a single grain size must be included. For simplicity
we shall refer to these additional grains as 'small', reserving the term
'normal' for the size estimated above from optical data. It might be
argued that a power-law distribution would be a good working hypothesis if
the small grains were thought to have the same composition as the normal
grains. (A power-law distribution has been measured for interplanetary
grains, but this may not be at all relevant to interstellar grains.) However, the shape of the ultraviolet extinction curve, especially the peak at
$4.6 \mu^{-1}$, might suggest an alternative concept of several distinct grain
species. One species would explain the optical data but would give relatively neutral extinction in the ultraviolet. A second species of much
smaller grains would explain the rapid rise in the far ultraviolet but
would contribute little at optical wavelengths. A second species of small
grains (thus a third species) might be needed to explain the prominent
$4.6-\mu^{-1}$ peak. Such a schematic three-species 'decomposition' of the
extinction curve is pictured in Fig. 3.3. Several different detailed ver-
sions have been proposed. More variants can be developed by introducing
core-mantle particles. The merits of such models may ultimately become
clear, but for the present they are plagued with severe problems of unique-
ness. To narrow down the range of possibilities the models must be subjected to data collected from a whole variety of observations, as will be
discussed in later chapters. Even then it would be prudent to regard the
derived model parameters with some scepticism.

How much dust is there? The amount of reddening in the Galactic
plane within a few kiloparsecs of the sun averages about

$$E'_{B-V} \simeq 0.6 \text{ mag kpc}^{-1} \qquad (3.1)$$

so that $A'_V \simeq 2$ mag kpc^{-1} when $R \simeq 3.3$. If we fall back on our simplistic

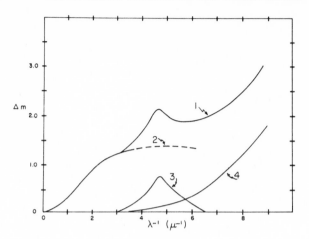

FIG. 3.3. A schematic decomposition of the extinction curve
(1) into three components (2,3,4). (From J.M. Greenberg and
S.S. Hong, The chemical composition and distribution of interstellar grains, in <u>Galactic radio astronomy</u> (eds. F.J. Kerr and
S.C. Simonson). Reidel, Dordrecht (1974).)

model of normal grains chosen to fit optical extinction then, using $Q_e(V) \approx$
2, we can find the number density of grains

$$n_d \approx 10^{-12}(a/10^{-5} \text{ cm})^{-2} \text{ cm}^{-3} \qquad (3.2)$$

and the mass density

$$\rho_d \approx 4 \times 10^{-27} (a/10^{-5} \text{ cm}) s_d \text{ g cm}^{-3} \qquad (3.3)$$

where s_d is the density of the grain material. By way of example, 'ice'
grains with $a \approx 0.3 \mu$ and $s_d \approx 1 \text{ g cm}^{-3}$ give $\rho_d \approx 10^{-26} \text{ g cm}^{-3}$, whereas
silicate grains with $a \approx 0.15 \mu$ and $s_d \approx 3 \text{ g cm}^{-3}$ would require $\rho \approx 2 \times$
$10^{-26} \text{ g cm}^{-3}$. The mass density of 'small' grains would be somewhat smal-
ler, by a factor roughly equal to the ratio of far-ultraviolet to optical
wavelengths ($\sim 1/3$).

To put these in context, we note that the total interstellar gas den-
sity is on average $\rho_g \approx 2 \times 10^{-24} \text{ g cm}^{-3}$, which corresponds to $n_H \approx 1 \text{ cm}^{-3}$.
There is generally a good correlation of gas and dust in space (§ 9.2).
Therefore we conclude that

$$\rho_d \approx 10^{-2} \rho_g . \qquad (3.4)$$

When it is recalled that elements other than hydrogen and helium amount to less than 2 per cent of the total interstellar mass (Table 10.1), it can be seen that a significant fraction of the 'heavy' elements must be tied up in the grains. In fact in some regions that have been studied it seems possible that all of the 'heavy' elements are used up, or at least that they are so depleted that further growth of grains is impossible (§9.3).

Because the above estimate of ρ_d/ρ_g is based on a specific and rather simple model, the question of its general validity should be examined. In this regard it can be observed that the results of a wide variety of detailed computations, in which there are different mixtures of particle composition and different distributions in size, mantle size, shape, and orientation, all point to the same conclusion. This apparent uniformity in the derived value of ρ_d may be understood and generalized using a powerful method based on the application of the Kramers–Kronig dispersion relations to the complex electric susceptibility of the interstellar medium (§10.1.1).

3.1.3. Extinction at X-ray wavelengths

Just beyond the Lyman limit photoelectric absorption by the interstellar gas makes observations of the properties of extinction by grains virtually impossible. However, at even shorter wavelengths, in the X-ray range, absorption by the gas is much reduced and X-ray sources are available as probes of the interstellar medium. The effects of absorption and scattering by grains can be considered separately. Calculations of the cross-sections can be carried out using the anomalous diffraction theory (eqn (2.36)) if the refractive index at X-ray wavelengths is known. For convenience we shall use the notation $m-1 = -\eta - i\beta$, where η and β are small numbers.

The absorption coefficient γ that is needed for Q_a can be found using the relation

$$\gamma = 4\pi\beta/\lambda = n_i \sigma_i \qquad\qquad\qquad (3.5)$$

where σ_i is the atomic photoelectric absorption cross-section for the species with number density n_i in the grain material. Tabulations of the results of theoretical and laboratory work generally refer to the parameter σ. It can be seen from eqn (2.38) that as long as the individual grain is not optically thick to the X-rays the amount of absorption by the grains cannot be distinguished from that by an equivalent mass of material in gaseous form. For any element, σ is largest at the wavelength of the absorption edge, decreasing towards shorter wavelengths. At the K edges of the important elements, C, N, and O, σ_K is typically less than 10^{-18} cm^2 (Table 3.1). Therefore in any material containing a mass fraction f of one

TABLE 3.1
SELECTED DATA FOR K ABSORPTION EDGES

Element	Z	λ_K(Å)	E_K(keV)	$\sigma_K(10^{-19}\text{cm}^2)$
C	6	43.6	0.284	11
N	7	31.0	0.400	(8.4)
O	8	23.3	0.532	6.4
Mg	12	9.51	1.30	2.2
Si	14	6.75	1.80	1.5
S	16	5.02	2.47	(1.2)

of these elements, $\gamma_K < 10^5$ f cm^{-1} at the K edge. For the grain sizes discussed above γa is probably less than unity, and so saturation would not be a serious problem ($Q_a < 1$).

A significant contribution to extinction also arises from scattering. Away from absorption edges this can be computed using the classical Lorentz result for scattering by the electrons, which can be considered unbound. Formally

$$\eta = 2.7 \times 10^{-6} \, s_d \, \lambda^2(\mathring{A}) \, Z/M \qquad\qquad (3.6)$$

where Z and M are the atomic charge and mass number respectively. Making use of the Rayleigh–Gans approximation (eqn (2.38))

$$Q_e \simeq Q_s \simeq 1.5 \times 10^{-4} \, s_d^2 \, (a/10^{-5} \, \text{cm})^2 \, \lambda^2(\mathring{A}) \ . \qquad\qquad (3.7)$$

At a wavelength of 1 Å (12 keV) Q_s is very small but by 10 Å it has risen sharply, typically to 0.1 if a $\simeq$ 0.2 µ. At even longer wavelengths η continues to rise, so that the approximate formula breaks down and the exact anomalous diffraction equations must be used. The simple expression (3.6) for η must also be adjusted to take into account the dispersion near the K edge; there η and β vary rapidly, and both make important contributions to the scattering cross–section. The characteristic decrease of Q_s near the K edge is illustrated in Fig. 3.4. (Note also the difference between the Rayleigh–Gans scattering approximation and the exact anomalous diffraction results.) The effect of this decrease in Q_s on the brightness of an X–ray halo around an X–ray source (§4.4.2) potentially could provide a means of identifying the constituent elements in the interstellar grains.

A measure of the amount of X–ray extinction to be expected can be obtained from the visual extinction through scaling based on the ratio of the values of Q_e. Since $A'_V \simeq$ 2 mag kpc^{-1} and $Q_V \simeq$ 2,

$$\tau'_X \simeq Q_X \text{ kpc}^{-1} \ . \qquad\qquad (3.8)$$

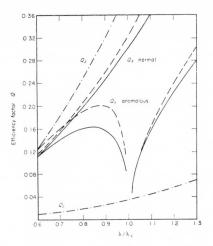

FIG. 3.4. The anomalous behaviour of Q_s near the K edge of
oxygen in ice grains of radius 0.2 μ (lower curves) compared
with the normal shape in the absence of the K edge (upper
curves). The dashed curves give the Rayleigh-Gans approximation. The corresponding variation of Q_a is also shown. (From
P.G. Martin, On the interaction of cosmic X-rays with interstellar grains, Mon. Not. R. astron. Soc. 149, 221 (1970).)

The energy range over which interesting effects could be observed is restricted at the high end to about 10 keV ($\simeq 10^4$ μ^{-1}) by low values of Q_X,
and at the low end to about 100 eV ($\simeq 100$ μ^{-1}) by a combination of the
saturation of Q_X and the increasing absorption by the interstellar gas.

3.2. LINEAR POLARIZATION

3.2.1. Observations of linear polarization

Single-colour optical observations of thousands of early-type stars have
been used to study linear polarization out to distances of several kiloparsecs in the Galactic disk. A typical degree of polarization (in the
visual) is 0.03 (3 per cent) over a pathlength of 1 kpc in the Galactic
plane. Considerable uniformity in position angle in some regions of the
sky has been found, suggesting grain alignment on a large scale (§3.2.3).

Additional multi-colour surveys of the wavelength dependence of the
linear polarization for several hundred stars have revealed a characteristic maximum in the visual, with a decrease to the red and ultraviolet.
Although the maxima for different stars occur at different λ_{max}, it has
been found that most observations are encompassed by the empirical formula

$$p(\lambda)/p_{max} = \exp\{-1.15 \ln^2(\lambda_{max}/\lambda)\} . \tag{3.9}$$

The parameter λ_{max} ranges from 0.45 to 0.80 µ, being typically 0.55 µ.

From region to region in the Galactic plane there is evidence for significant changes in the average value of λ_{max}. This is important if it points to similar differences in R, which should vary along with λ_{max} if extinction and polarization are caused by the same grains. To investigate this possibility, extinction curves from different regions could be plotted in a normalized manner using λ_{max} to scale the wavelength axis. Such a procedure in fact brings otherwise different-shaped extinction curves into better agreement, though there is often over-correction. The correlation of the changes in the wavelength dependences of linear and circular polarization is somewhat stronger (§3.3), as might be expected since the grains involved are more definitely the same.

Measurements in the infrared and in the ultraviolet are very rare and therefore present an observational enterprise of outstanding importance. The few infrared observations that exist appear to fall close to or slightly above the extrapolation of the empirical formula. However, the formula is based mainly on fitting optical observations and so extrapolation should be avoided, especially into the ultraviolet where we have seen evidence for an important additional component of 'small' grains.

One example of the data available is plotted in Fig. 3.5 for the star ζ Oph (HD 149757). The curve shown results from a fit to the optical data,

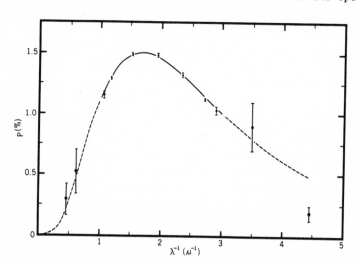

FIG. 3.5. The wavelength dependence of linear polarization of HD 149757 (ζ Oph). A curve of the form of eqn (3.9) has been fitted to the optical data. The broad coverage, into the infrared and the ultraviolet, is unusual.

giving p_{max} = 1.50 per cent and λ_{max} = 0.60 μ. It is evident that the
curve also satisfactorily describes the infrared data in this case. There
is a limited amount of higher resolution data in which small-scale structure can be sought. Results for HD 183143 reveal a smooth behaviour well
approximated by eqn (3.9).

3.2.2. Interpretation of linear polarization
The characteristic wavelength dependence observed bears a strong resem-
blence to the theoretical curves of ΔQ_e seen earlier (§2.4.2). Over the
restricted range of optical observations it is rather difficult to choose
between dielectric as opposed to metallic grains. The difference in the
behaviour predicted for ΔQ_e should, however, show up in the near-infrared
and ultraviolet when the observations become available. A dielectric composition seems able to satisfy the near-infrared and optical data that
exist. (It should always be remembered in interpreting these observations
that calculations based on a constant refractive index can be misleading if
used over such a wide wavelength range, especially for metallic grains.)
Fortunately the ambiguity in interpretation can be resolved using optical
observations of circular polarization (§3.3); the dielectric alternative
seems favoured.

   With these curves we can also investigate the grain properties
required to produce $\lambda_{max} \simeq 0.55$ μ (V filter). In the case of dielectric
grains, results for prolate spheroids show $\rho_{max} \simeq 1.8$, comparable with the
corresponding estimate (2.2) from extinction (§3.1.2). This raises the
question of whether all of the extinction is produced by the polarizing
grains. The answer is not straightforward because of several complications. For prolate spheroids ρ_{max} decreases when the particle shape is
more elongated. Values of ρ_{max} are generally higher for oblate particles,
decreasing when the particle shape becomes more spherical. In all cases
ρ_{max} increases with m-1. There are also shifts in the extinction curves Q_e
which partially compensate for the shifts in ρ_{max}. Computations of Q_e and
ΔQ_e would clearly help resolve this problem, but because they are expensive
detailed modelling of both extinction and polarization has received little
attention. In any case some assumption about the grain composition and
shape might have to be made. The possibility that most of the extinction
is produced by the polarizing grains does seem good; if not, then the problems of grain alignment become more severe.

   Observations of the ultraviolet polarization can also be used to
investigate the polarizing properties of the 'small' grains responsible for
the rising ultraviolet extinction. It seems that the points for ζ Oph
(Fig. 3.5) and another star κ Cas could be explained by the normal grains
alone. In particular there is no evidence for an excess in polarization at

the 4.6-μ^{-1} extinction peak. Small grains may therefore be spherical or
unaligned. Further observations through this peak and at even shorter wavelengths will clearly make a valuable contribution to understanding the
many possible components of the interstellar grains. The question of a
size or composition dependence of the degree of alignment is important in
determining the nature of the alignment mechanism itself.

Since grains must be aligned to produce the observed polarization,
the efficiency of the alignment process can be assessed using the ratio of
polarization to extinction. A plot of p_{max} against E_{B-V} for many stars
reveals a considerable scatter below an upper envelope with slope 9 per
cent mag^{-1}. Part of the scatter might originate in differences in the
alignment efficiency from region to region; in addition there may be depolarization if the position angle of the grain alignment varies along the
line of sight (eqn (2.21)). If we use $R \simeq 3.3$ and $\lambda_{max} \simeq 0.55~\mu$ then the
maximum to be explained is

$$(p/\tau)_V = \{(C_{eI}-C_{eII})/(C_{eI}+C_{eII})\}_V \simeq 0.028~. \tag{3.10}$$

The values of C_e to be used correspond to those in Fig. 2.1. To compute
these C_e´s an average over a distribution of orientations must be taken.
This distribution is determined by the alignment mechanism. Recall that
the Q_e values shown in Figs. 2.4-2.6 were calculated for a very special
static alignment which maximizes the elongation of the grain profile (picket-fence alignment). This type of alignment is very different from that
envisioned for tumbling interstellar dust particles.

In the latter case alignment is the combined result of two effects.
First, there must be a preferred orientation in space for the angular
momentum vector $\underline{J}$ of the grain. Later we shall discuss why this preferred
direction is thought to be along the interstellar magnetic field (§3.2.3)
and how magnetic alignment might be caused (§8.4). The degree of orienta-
tion of $\underline{J}$ with respect to the preferred direction, say $\underline{B}$, will be described
by

$$q_{JB} = \langle P_2(\cos \theta_{JB})\rangle \tag{3.11}$$

where θ_{JB} ($<90°$) is the angle between the field direction and $\underline{J}$. Note that
even spherical particles might have significant alignment of $\underline{J}$, but of
course no polarization would arise. The second effect to be considered
therefore is the alignment of $\underline{J}$ with respect to the body axes of a non-
spherical grain. In thermal equilibrium, in which equipartition of rotational energy may be assumed, a grain tends to rotate around a short axis
with a high moment of inertia. For an oblate spheroid the tendency is

towards the symmetry axis $\underline{S}$, but for a prolate spheroid it is perpendicular
to $\underline{S}$. In either case a measure of the alignment of $\underline{S}$ and $\underline{J}$ is q_{SJ}, defined
in the same manner as q_{JB}. If in addition the direction of $\underline{B}$ fluctuates
around a mean $\tilde{B}$, then an additional factor $q_{B\tilde{B}}$ can be defined. What we are
interested in ultimately is the alignment of $\underline{S}$ with respect to $\tilde{B}$, which can
be described by $q_{S\tilde{B}}$. When $\underline{J}$, $\underline{S}$, and $\underline{B}$ have rotational symmetry around $\underline{B}$,
$\underline{J}$, and $\tilde{B}$ respectively, then

$$q_{S\tilde{B}} = f q_{SJ} q_{JB} q_{B\tilde{B}} \qquad\qquad (3.12)$$

where the factor f, near unity, is needed to compensate for some correlation of the angles for partially aligned grains. This shows how the
effects must combine to give alignment. It is unfortunately often the case
that the alignment of $\underline{S}$ and $\underline{J}$ is low even when that of $\underline{J}$ and $\underline{B}$ is significant, and so the net alignment is small. Note that the electric vector is
either along or perpendicular to the projection of $\underline{B}$ on the sky.

    There is no general analytical expression relating the polarization
and extinction arising from incomplete alignment to that predicted for
picket-fence alignment. However, an approximate relation may be developed
using the assumption that the dependence of the cross-sections on inclination is as described by the simple formulae in the long-wavelength approximation (eqn (2.34)). The result is that

$$C_{eI} + C_{eII} \simeq 2\{(2C_{e1} + C_{e2})_{pf}/3\}\{1 + O(p/\tau)\} , \qquad\qquad (3.13)$$

$$C_{eI} - C_{eII} \simeq q_{S\tilde{B}} (C_{e1} - C_{e2})_{pf} \sin^2\psi \qquad\qquad (3.14)$$

where ψ is now the angle between $\tilde{B}$ and the line of sight, and on the
right-hand side the C_e's are those calculated for picket-fence alignment.
In interstellar applications the term of order p/τ can be neglected. The
dominant term in the extinction could be replaced by twice the average of
the three principal cross-sections (replace one C_{e1} term by the 'end-on'
cross-section) with little effect on the estimate at visual wavelengths.
Actual numerical computations show that these analytical relationships are
fairly good, and so we use the approximation

$$(p/\tau)_V \simeq q_{S\tilde{B}} \sin^2\psi \; (p/\tau)_{pf,V} \simeq 0.028 \qquad\qquad (3.15)$$

to estimate $q_{S\tilde{B}}$. Conversely, the maximum polarization for any ψ is produced by perfect spinning alignment, with $q_{S\tilde{B}}=1$ for oblate particles and
$q_{S\tilde{B}}=-0.5$ for prolate particles. Note that the same sense of polarization
is produced in each case because the sign difference in $q_{S\tilde{B}}$ is cancelled by

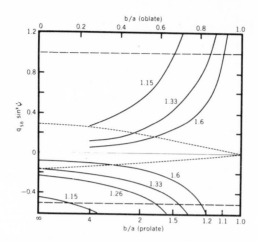

FIG. 3.6. Values of the alignment parameter required to produce the maximum observed ratio of linear polarization to
extinction. Curves are for homogeneous spheroids with real
refractive indices as labelled. Long-dash lines give the maximum values possible for any grain; short-dash lines give the
limiting values reached in maximal Davis-Greenstein alignment
($8.4).

the sign difference in ΔC_e.

Some curves of $(p/\tau)_V$ have already been given in Figs. 2.4-2.6 where
the dependence on axial ratio and $m-1$ for dielectric particles was seen.
If we identify the V passband with ρ_{max} then the values of $q_{S\tilde{B}} \sin^2\psi$
required to satisfy eqn (3.15) can be found. Loci of such values are shown
in Fig. 3.6. Valuable restrictions on the possible combinations of axial
ratio and $m-1$ are obtained from the intersections of the loci with the
extrema mentioned above (1 and -0.5). The 'allowed' region for $\sin^2\psi=1$ is
indicated in Fig. 3.7. More stringent limits would apply if the polarizing
grains produce only a fraction of the visual extinction, if $\sin^2\psi < 1$, or
if the alignment is far from perfect, as seems likely ($8.4). The differ-
ences between oblate and prolate particles should not be over-interpreted,
since under the same physical conditions $q_{S\tilde{B}}$ may be somewhat smaller for
oblate than for prolate particles.

3.2.3. Linear polarization as a tracer of the Galactic magnetic field
In §8.2 we shall see that precession of the angular momentum vector makes
magnetic alignment probable in even a weak interstellar magnetic field. If
the direction of symmetry of grain alignment is provided by the Galactic
magnetic field then systematic observations of optical polarization (position angle and strength of the electric or E vector) as a function of

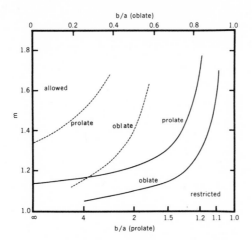

FIG. 3.7. Grains having combinations of m and axial ratio
below the solid loci shown cannot account for the maximum ratio
of polarization to extinction even with optimum alignment (see
Fig. 3.6). With maximal Davis–Greenstein alignment the allowed
parameters are even more restricted, as indicated by the broken
curves.

direction and distance can be used to map out the field. Only the projection of the field on the plane of the sky is seen. Also note the two
alternative orientations of the E vector with respect to the field, parallel and perpendicular. Alignment of grains by means of paramagnetic
absorption (§8.4) predicts that the E vectors should be parallel to the
magnetic field. We shall examine the evidence for magnetic alignment.

Optical polarization observations of several thousand stars have now
been accumulated. One useful way to illustrate some of the basic information is with E-vector maps, such as in Fig. 3.8. The polarization for each
star is represented by a bar giving the degree of polarization and the
orientation of the E vector. Notice that the strength of polarization and
the pattern of E vectors for the distant stars is not that much different
than for stars ten times closer. (The lack of data at high b for large
distances results from the concentration of B stars, those surveyed, in the
Galactic plane.) Optical polarization data therefore reflect relatively
local conditions.

The second point, which has been recognized since early investigations, is that in certain directions there is rather uniform alignment parallel to the Galactic plane. The direction $\ell \simeq 140°$ is commonly cited; the
opposite direction, $\ell \simeq 320°$, is also uniform. It is argued that the
required large-scale (>100 pc) order could be provided by a magnetic field
lying most plausibly in the Galactic plane and running perpendicular to the

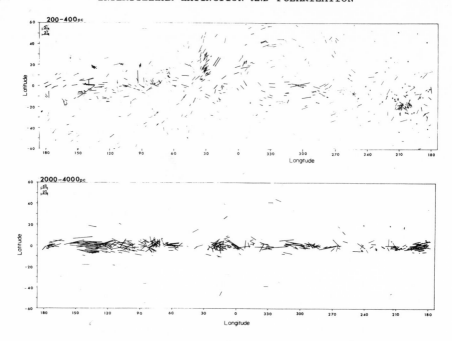

FIG. 3.8. Maps of electric vectors of optical linear polarization for two different distance ranges. The length of the bar
measures the degree of polarization. Notice the regions of
uniform alignment. (From D.S. Mathewson and V.L. Ford, Polarization observations of 1800 stars, Mem. R. astron. Soc. 74,
139 (1970).)

directions of maximum alignment. Hence the E vectors observed would be
parallel to the Galactic field. The disorder in E vectors seen at other
longitudes is also consistent with a longitudinal field directed locally
towards $\ell \simeq 50°$. In this direction no polarization would be observed in a
uniform field; what is seen must be the result of a fluctuating component
of the field and is therefore random in position angle. The strength of
the fluctuating component is apparently comparable with the mean. Fluctuations are also present along lines of sight perpendicular to the field, but
are less apparent in the maps owing to the non-zero mean. The above statements about the direction of the field can be quantified somewhat by computing averaged properties of polarization as a function of galactic longitude. For example it is found that the minima in dispersion of position
angle occur at $\ell \simeq 135°$ and $310°$, and that the minima in polarization occur
at $\ell \simeq 50°$ and $230°$, in agreement with what is inferred from the maps.
Added interest is provided by the coincidence of the longitudinal direction
of the aligning force with the direction of the local spiral feature (the
Orion arm) delineated by optical tracers.

There are other observations which support this circumstantial evidence for the orientation of the E vectors and the topology of the field.
Analysis of the rotation-measures of extragalactic radio sources and pul-
sars suggest that the mean field (as detected now through the line-of-sight
component) lies in the Galactic plane, but at a somewhat greater longitude,
$\ell \simeq 90°$, that is closer to the spiral arm direction indicated by H I measures. Since larger pathlengths are involved in these measurements, the
rotation-measure data can be made consistent with the optical data if there
is a local bending of the field direction due to some perturbation, perhaps
from a magnetic-bubble instability or a supernova explosion. A satisfactory explanation must await a comprehensive analysis of the field structure.

More direct evidence is provided by measurements of the polarization
of the background synchrotron radiation at radio wavelengths. The radio
polarization vectors are perpendicular to the field and map out the tangential component. Because of the effects of Faraday depolarization the
polarization seen must be of fairly local origin. In the $\ell \simeq 140°$ region
the polarization is oriented perpendicular to the Galactic plane and is
particularly strong, indicating a relatively uniform field. This provides
perhaps the best direct evidence for the optical E vectors lying along the
field. While similar coincidences are found elsewhere, in loops of polarized emission, there are some apparent counter-examples. For instance the
ridge of radio continuum emission called the North Polar Spur (the northern
extension of Loop I) is traced out fairly well by optical E vectors. It
can be seen arching up towards lower ℓ from $\ell \simeq 30°$, $b \simeq 0°$ in the data for
nearby stars (Fig. 3.8), showing it to be a local feature (it is obvious
even in the 100–200 pc range). The magnetic field determined by radio continuum measurements runs along the ridge for $b > 40°$, thus being consistent
with optical E vectors lying along the field, but is actually perpendicular
to the ridge at lower latitudes. This geometry might be explained with a
model of the expansion of a nearby supernova remnant into the interstellar
medium; the discrepancy in field orientation would be resolved if the syn-
chrotron-emitting relativistic electrons and the polarizing grains were not
coincident in space. The overall picture is therefore not without ambiguity.

An interesting additional point concerns the elongation of (optical)
dust clouds and (21-cm) H I clouds. For each a correlation of the direction of elongation with the orientation of optical E vectors has been
demonstrated. If the E vectors are taken as tracers of the field then the
question is why the clouds are elongated along the field. This is not
necessarily what is expected; for example in collapsing clouds gas motions
are restricted to parallel to the field even with low fractional ioniza-

tion, and so sheet-like configurations might arise leading to the opposite
correlation. The observed cloud structure may be more closely related to
the mechanism of confinement, or possibly some (magnetic) instability is
responsible. A similar and perhaps related phenomenon has been discovered
in some reflection nebulae, where E vectors of transmitted light are
aligned along fine filamentary structure (§5.3).

3.3. CIRCULAR POLARIZATION

The grains that cause the linear polarization just described also make the
interstellar medium linearly birefringent. The medium is then able to
transform linear polarization into elliptical polarization, just as a weak
waveplate does. The efficiency of the linear-to-circular conversion is
low, so that interstellar circular polarization is very small ($V/I \simeq 10^{-4}$)
and consequently difficult to detect.

The linear polarization which is to be transformed might be intrinsic
to the source. Specifically, the variations in the position angle of the
polarized synchrotron radiation across the face of the Crab Nebula have
been used to demonstrate the modulation and sign changes of circular polarization characteristic of a waveplate. Alternatively, the linear polarization may originate in the interstellar medium itself, being subsequently
transformed if the position angle of the grain orientation changes along
the line of sight. (Recall the simple two-slab model in §2.1.3.) Circular
polarization produced in this manner, called interstellar circular polarization, is much more common, allowing measurements in many more directions
in the Galaxy.

Fig. 3.9 displays measurements of the linear and circular polarization of HD 204827, both of which are relatively strong. The circular
polarization changes sign from the red to the blue spectral region. This
sign change is found to be characteristic of all measurements of interstellar birefringence. Closer examination of a number of stars with a range in
the value of λ_{max} reveals a correlated variation of λ_c, as expected, and
that $\lambda_c \simeq \lambda_{max}$. It was concluded from the curves computed for constant m
in §2.4.2 that as n′ increases relative to n-1 the ratio λ_c/λ_{max} progres-
sively increases from a value near unity when n′=0. The observations are
therefore consistent with dielectric grains (like ice and silicate), but
not with 'metallic' grains (like graphite) for which n′ is consistently
high. Not too fine a distinction between pure dielectrics and 'dirty' dielectrics (which absorb as well as scatter) can be made without performing
additional computations for realistic refractive indices which may vary
with wavelength. In fact for each material whose refractive index varies
rapidly over the optical spectrum (e.g. magnetite) it may be necessary to

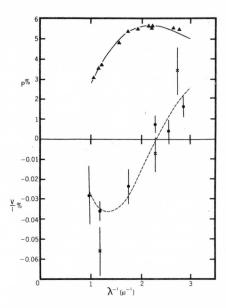

FIG 3.9. Wavelength dependences of linear and circular polarization for HD 204827. Circular polarization changes sign at a
wavelength near the peak in linear polarization.

carry out a separate calculation to see if a match to both the observed
linear and circular wavelength dependences can be obtained.

Recall that the degree of circular polarization of starlight depends
not only on the strength of the birefringence and linear polarization, but
also on the unknown amount of twist in the grain alignment along the line
of sight, through the parameter G (eqn (2.25)). Therefore the Crab Nebula
provides a unique opportunity to examine whether the level of birefringence
coincides with that expected on the basis of the known interstellar linear
polarization along the same path. The observed ratio $\Delta\epsilon/\Delta\sigma$ is found to
agree with the prediction from the dielectric grain models and the sign is
correct, adding confidence to this interpretation. (Note that deviations
from perfect spinning alignment cause similar reductions in birefringence
and dichroism.) With this background, model calculations can be used to
analyse the observations of stars to find the geometry-dependent factor G.

Because model computations are time consuming it is useful to develop
an analytic formula like that found for linear polarization (eqn (3.9)).
The Kramers–Kronig relations can be applied to the complex refractive index
of the interstellar medium (§10.1) for each of directions I and II in Fig.
2.1 to show quite generally that

$$b(f) = 2\pi^{-1} \int_0^1 \{d(fy)-d(f/y)\}/(y^2-1)\, dy \qquad (3.16)$$

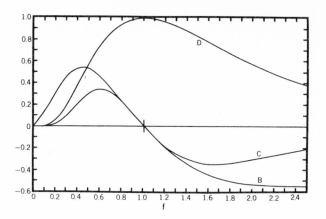

FIG. 3.10. Empirical formulae for the reduction of polarization observations. Abscissa $f=\lambda_{max}/\lambda$. Curve D is from eqn
(3.9), B is from eqn (3.16) with $B_o=0$, and C=DB. Their use
outside the optical range is not advisable. (From P.G. Martin,
A semiempirical formula for interstellar birefringence, <u>Astro-
phys. J</u>. 201, 373 (1975). University of Chicago Press. ©1975
by the American Astronomical Society. All rights reserved.)

where b(f) is the birefringence corresponding to the dichroism d(f) and f =
λ_{max}/λ. To evaluate the integral, d(f) must be known at all wavelengths,
which is not the case in the interstellar medium. However, we can try the
substitution of D(f), the empirical formula found to hold at optical wavelengths. Call the resulting expression B(f). The integrand is finite
through [0, 1] and is readily evaluated numerically using a simple scheme
like Gaussian quadrature. Notice that since $D(f) = D(f^{-1})$, then B(f) =
$-B(f^{-1})$ and B(1) = 0, or $\lambda_c = \lambda_{max}$. These curves are shown in Fig. 3.10.
We can also consider what would happen if the actual d(f) were, say, larger
than D(f) in the unobserved infrared region. Then the characteristic dispersion shape of b(f) would be preserved because d(f) is peaked near f=1,
but the whole curve would be displaced downwards causing λ_c to exceed λ_{max}.
To allow for this type of possibility the semi-empirical formula b(f) ≃
B(f) + B_o can be tried. A suitable expression must of course fit the
observations and must adequately describe the model calculations for a variety of possible refractive indices. Fig. 3.11 shows the reasonable agreement possible with this formula. As was mentioned above, the few stars for
which the wavelength dependence of circular polarization has been studied
in any detail suggest that $B_o \simeq 0$.

Quite generally the degree of circular polarization is

$$q(f) \simeq b(f)\ d(f)\ p_{max}^2\ G\ .$$ (3.17)

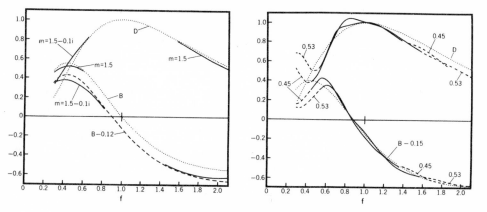

FIG. 3.11. (Left) Curves D and B (broken curves) have been
fitted to grain-model calculations (infinite cylinders) for
constant refractive indices m=1.5 and m=1.5-0.1i. Only the
deviations of the model calculations from the approximations
are shown. Note that the birefringence of the absorbing material changes sign at f<1. (Right) Same comparison for magnetite
with sizes giving λ_{max}=0.45 and 0.53 μ. The curves are broken
for λ>1 μ because the adopted refractive index becomes unrelia-
ble and for λ<0.33 μ as a reminder that very few observations
have been made in this region. (From P.G. Martin, A semiempirical formula for interstellar birefringence, Astrophys. J.
201, 373 (1975). University of Chicago Press. ©1975 by the
American Astronomical Society. All rights reserved.)

Using D(f) and the above approximation for b(f), eqn (3.17) can be fitted
to the circular polarization data of any star (even a single-colour observation) to find G, assuming that p_{max} and λ_{max} have already been determined
from linear polarization. For example for HD 204827, λ_{max} = 0.46 μ, p_{max} =
5.6 per cent (linear), and λ_c = 0.44 μ, G = -0.32 (circular). Smooth
curves fitted to the data are shown in Fig. 3.9. This value of G corresponds to a change in alignment by φ = +30 in a model with two equal slabs
(eqn (2.25) with r=1). Note that the derived sign is imporant as it gives
the direction of the twist.

Finally it may be remarked that extension of circular polarization
into the infrared and ultraviolet should provide valuable new restrictions
concerning the questions of grain composition and alignment.

FURTHER READING

§3.1.1

HACKWELL, R.D., and GEHRZ, R.D. (1974). Infrared photometry of high-lumi-
nosity supergiants earlier than M and the interstellar extinction
law. Astrophys. J. 194, 49.

HERBST, W. (1975). R-associations II. The ratio of total to selective
extinction. Astron. J. 80, 498.

----. (1978). Extinction law in dust clouds and the young southern cluster
NGC 6250: Further evidence for high values of R. Astron. J. 82,
902.

SAVAGE, B.D. (1975). Ultraviolet photometry from the orbiting astronomical
observatory XX. The ultraviolet extinction bump. Astrophys. J.
199, 92.

YORK, D.G., DRAKE, J.F., JENKINS, E.B., MORTON, D.C., ROGERSON, J.B., and
SPITZER, L. (1973). Spectrophotometric results from the Copernicus
satellite. VI. Extinction by grains at wavelengths between 1200 and
1000 Å. Astrophys. J. Lett. 182, L1.

SCHILD, R. (1977). Interstellar reddening law. Astron. J. 82, 337.

§3.2.1

SERKOWSKI, K., MATHEWSON, D.S., and FORD, V.L. (1975). Wavelength dependence of interstellar polarization and ratio of total to selective
extinction. Astrophys. J. 196, 261.

GEHRELS, T. (1974). Wavelength dependence of polarization. XXVII. Interstellar polarization from 0.22 to 2.2 μm. Astron. J. 79, 590.

§3.2.2

PURCELL, E.M. (1969). On the alignment of interstellar dust. Physica
41, 100.

§3.2.3

HEILES, C. (1976). The interstellar magnetic field. A. Rev. Astron.
Astrophys. 14, 1.

§3.3

MARTIN, P.G., and ANGEL, J.R.P. (1975). Systematic variations in the wavelength dependence of interstellar circular polarization. Astrophys.
J. 207, 125.

4

SCATTERING OF RADIATION

4.1. SCATTERING OF POLARIZED RADIATION

For a given state of polarization of the incident radiation and orientation
of the particle, the intensity and polarization of the scattered radiation
are to be calculated. The state of polarization is known when the amplitude functions for two orthogonal orientations of the electric vector are
specified. These directions, which will be called 1 and 2, are usually
oriented such that a right-handed triad is completed by the direction of
propagation, axis 3, and direction 1 is perpendicular to a chosen plane of
reference which contains axis 3. The terminology TE and TM or r and 1
(perpendicular and parallel) is often used to identify the modes 1 and 2
respectively.

There may be a different reference plane for each of the incident and
scattered radiation fields. In the case of a symmetrical particle each
reference plane is often chosen to contain the symmetry axis and the propagation vector. The particular conventions which will be used below are
illustrated in Fig. 4.1. Axis z is the symmetry axis. Incident radiation
propagates along axis 3 in the x-z plane at an incident angle ζ measured
from z. Axis 1 is chosen to lie along the negative y axis. For the scattered radiation, axis 3 is located by the azimuthal angle φ (measured from

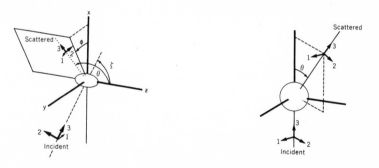

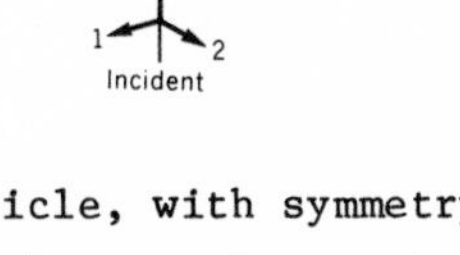

FIG. 4.1. Scattering by a spheroidal particle, with symmetry
axis z. Incident radiation propagates in the x-z plane, at
incident angle ζ. Angle θ locates the scattered radiation in a
plane inclined at angle φ.

FIG. 4.2. Scattering by a sphere. Scattering angle θ is measured in the scattering plane defined by the incoming and outgoing propagation vectors.

x through y) and the polar angle θ (measured from z). Axis 1 is in the direction of increasing ϕ (as for the incident radiation). Note that the symmetrical-particle co-ordinate system used above in the treatment of transmission is consistent with the choice here.

In the special case of a spherical particle it is convenient to take both reference planes to be the plane of scattering, which for any particle is defined to be that which contains both the propagation vectors. A single angle, say θ, measured between the propagation vectors is sufficient to describe the scattering. (The orientation of the scattering plane is also needed in many applications.) Such a scheme is illustrated in Fig. 4.2.

With this in mind a matrix of amplitude functions T_{ij} which relate the scattered to the incident amplitudes can be defined as (see eqn (2.3))

$$\begin{pmatrix} E_1 \\ E_2 \end{pmatrix}_s = -i \ \exp(-ikr)/kr \begin{pmatrix} T_{11} & T_{21} \\ T_{12} & T_{22} \end{pmatrix} \begin{pmatrix} E_1 \\ E_2 \end{pmatrix} . \tag{4.1}$$

(Note the unusual ordering of the subscripts.) The T's are functions of the scattering angle and depend also on the choice of reference planes. The off-diagonal terms are generally not zero, so that there is a mixing of the polarization modes (cross-polarization terms).

The scattering cross-section C_s is defined as the area that would intercept an amount of energy from the incident plane wave equal to the energy actually scattered in all directions. Therefore, from eqn (4.1), we have for the four combinations of the two pairs of modes four partial cross-sections

$$C_{sij} = k^{-2} \int |T_{ij}|^2 \ d\Omega . \tag{4.2}$$

Since the integral is over all directions of the scattered radiation for a fixed polarization of the incident wave, it is not appropriate to make use of axis labels with respect to the scattering plane (Fig. 4.2) in this particular formulation. The sums $C_{si} = C_{si1} + C_{si2}$ (i=1,2) which describe the total scattering of the two incident modes, are those most commonly calculated. The total cross-section for natural light is $C_s = (C_{s1} + C_{s2})/2$, similar to the expression for extinction.

To describe the pattern of scattered radiation as a function of direction, either the dimensionless intensity functions $|T_{ij}|^2$ or the differential cross-sections $dC_{sij}/d\Omega$ can be used. As another alternative a phase function can be defined as

$$P_{ij} = 4\pi \ C_{sij}^{-1} \ dC_{sij}/d\Omega ; \tag{4.3}$$

note the particular normalization of p.

In order to describe the polarization of the scattered wave, eqn (4.1) can be recast into a transformation equation for the Stokes vector of the form

$$\underline{W}_s = r^{-2} \, \underline{\underline{F}} \, \underline{W} \, . \tag{4.4}$$

The conventions for the Stokes parameters adopted here are exactly those used for the transmitted radiation in §2.1.2. Thus, for example, positive Q corresponds to the electric vector lying in the reference plane and position angles are measured in the usual astronomical sense. The resulting matrix

$$\underline{\underline{F}} = \begin{pmatrix} \frac{1}{2}(M_2+M_3+M_4+M_1) & \frac{1}{2}(M_2-M_3+M_4-M_1) & -S_{23}-S_{41} & -D_{23}-D_{41} \\ \frac{1}{2}(M_2+M_3-M_4-M_1) & \frac{1}{2}(M_2-M_3-M_4+M_1) & -S_{23}+S_{41} & -D_{23}+D_{41} \\ -S_{24}-S_{31} & -S_{24}+S_{31} & S_{21}+S_{34} & D_{21}-D_{34} \\ D_{24}+D_{31} & D_{24}-D_{31} & -D_{21}-D_{34} & S_{21}-S_{34} \end{pmatrix} \tag{4.5}$$

where

$$S_{ij} = \mathrm{Re}(A_i A_j^*) \, , \qquad D_{ij} = -\mathrm{Im}(A_i A_j^*) \, , \qquad M_i = S_{ii} = A_i A_i^* \tag{4.6}$$

and for compactness T_{11}, T_{22}, T_{12}, and T_{21} have been replaced by $k(A_1, A_2, A_3,$ and $A_4)$ respectively. The asterisk denotes the complex conjugate. Note that the M's are the differential cross-sections described above, and that the element F_{11} is the differential cross-section for natural light. The corresponding phase function $4\pi F_{11}/C_s$ is frequently used. The other components F_{i1} in the first column are also quite important as they give rise to polarization of the scattered wave even when the incident radiation is unpolarized. Many simplifications of this matrix occur for symmetrical particles and particular orientations of the propagation vectors. Scattering by a sphere is considered in some detail below.

4.2. CALCULATIONS FOR SINGLE PARTICLES

4.2.1. Differential equation formulation

It will be recalled that in the process of deriving the extinction cross-section (§2.2.1) the scattering amplitude (for any direction) was expressed as an infinite series in the appropriate angular wavefunctions. The expansion coefficients were found from matching conditions at the particle boundary. Only the forward-scattering amplitude was used. However, the only extra computational effort to find the T_{ij}'s for any scattering angle is evaluation of the angular wavefunctions. Differential scattering cross-sections, phase functions, and matrices like $\underline{F}$ follow simply from the above formulae.

As an example of how this appears in practice consider a sphere for which $T_{12} = T_{21} = 0$ and (see eqn (2.27))

$$T_{11}(\theta) = \sum_{n=1}^{\infty} \{(2n+1)/n(n+1)\}\{a_n \pi_n(\cos\theta) + b_n \tau_n(\cos\theta)\} \qquad (4.7)$$

$$T_{22}(\theta) = \sum_{n=1}^{\infty} \{(2n+1)/n(n+1)\}\{b_n \pi_n(\cos\theta) + a_n \tau_n(\cos\theta)\} \qquad (4.8)$$

where θ is the scattering angle in the notation of Fig. 4.2. The angular wavefunctions were given above in eqn (2.28). The scattering cross-sections are obtained by integrating the differential cross-sections. This can be done in two ways: either the differential cross-section is evaluated and integrated numerically, or it is integrated analytically and then evaluated. In the case of a sphere where the latter is possible, $C_s = C_{s1} = C_{s2}$ and

$$C_s = 2\pi k^{-2} \sum_{n=1}^{\infty} (2n+1) (|a_n|^2 + |b_n|^2) . \qquad (4.9)$$

Other shapes for which this has been carried out are those previously described: prolate and oblate spheroids, coated spheres, and infinite circular cylinders. The latter is a rather special case, since the scattered radiation is confined to a conical surface. The cone has its axis parallel to the cylinder axis and has a full opening angle 2ζ, where ζ is the angle of incidence measured from the cylinder axis. For perpendicular incidence ($\zeta = 90°$) the conical surface becomes the plane perpendicular to the cylinder axis. Despite this peculiarity the analysis is much the same.

4.2.2. Alternative methods
In the integral equation formulation it is necessary to solve the coupled set of linear equations for each different angle of incidence and mode of polarization. Not much additional computation is required to specify the scattering amplitudes for more than one outgoing direction. (The same applies to the relative amounts of effort expended in the differential equation technique.) Therefore the requirements for studying the phase functions and matrix $\underline{F}$ over the wider range of particle shapes made possible by the technique are not too restrictive. No study of total scattering cross-sections has been made by this method, but there should be no fundamental difficulties.

The discrete dipole array approximation (§2.2.3) is also useful for scattering calculations, since the far-field scattering amplitudes of the dipoles may be summed for any direction and electric vector orientation. All of the related quantities such as the intensity functions and the transformation matrix $\underline{F}$ can then be found. Numerical integration of the differential cross-section over all solid angles (eqn (4.2)) gives the total scattering cross-section. The latter can also be computed independently as the difference of the extinction and absorption cross-sections (§2.2.3), providing a valuable check on the accuracy of this approximation.

The application of the microwave analogue method to scattering is in many ways similar, with the exception that the complex scattering amplitudes have to be measured rather than calculated. There have not been many experiments of this type, most efforts being concentrated on extinction cross-sections. In fact no calculations of the total scattering cross-section (and hence the absorption cross-section) by this technique have been presented.

4.3. INPUT FOR RADIATIVE TRANSFER

Basic elements required for any calculation of radiative transfer in a dusty environment are the total scattering cross-section C_s, the (normalized) phase function, and the albedo C_s/C_e. The latter is one way of specifying the relative contributions of scattering and absorption to the total optical depth. A variation in any of these three elements affects the end result. One fundamental way in which variations arise, even for a fixed arrangement of grains, is through the wavelength dependence of the basic elements. Thus observations of a scattering phenomenon as a function of wavelength have the potential of separating the effects characteristic of the grains from those characteristic of the geometry. Some features of the dependence of the basic elements on wavelength (and size and refractive index) will be illustrated. Discussion of certain approximations is deferred to §4.4.

Scattering cross-section. Efficiency factors for a sphere are shown in Fig. 4.3 as a function of the parameter x which embodies scaling of a and λ. The upper curve is for a particle material with real refractive index (m=1.33). In this special case $Q_s = Q_e$ and all of the extinction is accounted for by scattering. Comparison of independent evaluations of Q_s and Q_e provides a useful check on numerical accuracy. For two absorbing materials (m=1.33-0.1i and m=1.33-0.33i) the relationship between Q_s and Q_a (the efficiency factor for absorption) can be seen (cf. Fig. 2.3 for Q_e). Note how the presence of absorption has damped out the effects of interference seen above in Q_s. When inspecting these figures recall the earlier identification of x≈3 (n'=0) or x<3 (n'>0) with the optical region of the

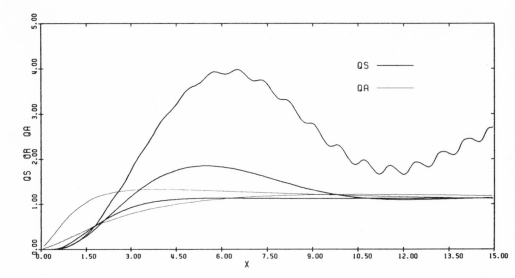

FIG. 4.3. Efficiency factors for scattering and absorption by
spheres as a function of the size parameter x. The three
curves for Q_s are for m=1.33, m=1.33-0.1i, and m=1.33-0.33i.
For m=1.33, $Q_s=Q_e$ and $Q_a=0$. Introduction of absorption damps
out the large peak in Q_s and increases Q_a. For Q_e see Fig.
2.3.

spectrum.

    Albedo. The ratio of the scattering to the extinction cross-section
is called the albedo $\tilde{\omega}$. Two limiting cases are pure scattering ($\tilde{\omega}=1$) and
pure absorption ($\tilde{\omega}=0$). The behaviour of the albedo for m=1.33-0.1i and
m=1.33-0.33i is illustrated in Fig. 4.4. Note that for absorbing particles
the albedo is not constant even if the refractive index is independent of
wavelength. At long wavelengths (small x) absorption dominates, so that
$\tilde{\omega}\approx0$ (see §4.4.1). At the other extreme $\tilde{\omega}\approx0.5$ since $Q_s\approx Q_a\approx1$ (see §4.4.3 for
the exact relation).

    Phase function. Other than the individual intensity functions $|T_{ij}|^2$
which are sometimes displayed, the most commonly plotted quantity is the
phase function for natural light $(4\pi F_{11}/C_s)$. Some of the important properties of the phase function are illustrated in Fig. 4.5 (left).
    For small x the phase function is relatively flat; it is called the
Rayleigh phase function for natural light (see §4.4.1). Note that the
scattering is not isotropic ($p\neq1$). As x becomes larger the phase function
develops an increasingly sharp forward peak, which in the extreme is identified with the diffraction phenomenon. Interference effects produce

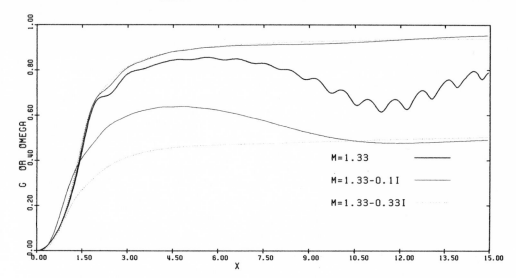

FIG. 4.4. The x dependence of g and $\tilde{\omega}$ for spheres with three different refractive indices. The three curves ending at the upper right represent g. For m=1.33, $\tilde{\omega}$=1.

oscillations and ripples which again are of little relevance in astronomical applications since they are averaged out in a size distribution. Accurate averaging of computed phase functions by numerical integration requires considerable care. Similar curves for absorbing spheres are shown in the lower part of Fig. 4.5. Recall how Q_e and Q_s are also affected by absorption (Figs. 2.3 and 4.3). It is rather difficult to generalize since the details are very much dependent on the size, wavelength, and refractive index (through x and m). Note that ρ is not a suitable scaling parameter for phase functions.

The degree of forward scattering is very important in radiative transport problems. For example, when the phase function is strongly peaked, radiation can penetrate to a greater optical depth before being absorbed; the effective albedo is higher than it is for more isotropic scattering. For simplicity spherical particles are used in many scattering computations. Even then it is often preferable to work with an analytic expression for the phase function. One approach used is to expand the phase function as a series of Legendre polynomials:

$$p(\theta) = 1 + \sum_{n=1}^{\infty} (2n+1)\, p_n\, P_n(\cos\theta) \, . \tag{4.10}$$

The coefficients p_n can be evaluated directly from the Mie coefficients. Alternatively, for single particles or ensembles of particles, they can be obtained routinely by a least-squares fit.

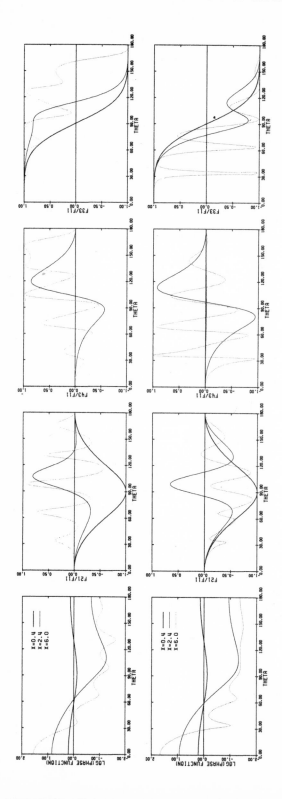

FIG. 4.5. Elements of the transformation matrix for scattering by a sphere. Recall that θ=0°is the forward direction. The upper panel is for m=1.33; values corresponding to the cases x=0.4, 2.4, 6.0 are Q_S =0.006, 1.13, 3.89 and g=0.03, 0.69, 0.85 (ω̃=1). The lower panel is for m=1.33-0.331, with Q_S=0.006, 0.76, 1.13, g=0.03, 0.75, 0.91, and ω̃=0.02, 0.38, 0.47. Curves for x=0.4 are close to the approximations in the long-wavelength limit. On the left is the logarithm of the phase function for natural light, showing strong forward scattering as x increases. The degree of linear polarization from single scattering is shown next; it is largest for small x. (Rapid oscillations will be averaged out in a size distribution.) On the other hand, the amount of linear-to-cir-cular conversion is least for small x (for x=0.4 it is imperceptibly small in these figures). For completeness the diagonal element which converts U to U and V to V is also shown on the right.

The coefficient p_1 is often called the asymmetry parameter and denoted g. Generally

$$g = <\cos \theta> = \tfrac{1}{2} \int_0^\pi \cos \theta \, p(\theta) \sin \theta \, d\theta \, . \tag{4.11}$$

It can be evaluated numerically from this formula for any phase function, but for a single sphere it is, from the Mie theory,

$$g = (4\pi/k^2 C_s) \, [\sum_{n=1}^\infty \{n(n+2)/(n+1)\} \, \text{Re}(a_n a^*_{n+1} + b_n b^*_{n+1}) +$$
$$\sum_{n=1}^\infty \{(2n+1)/n(n+1)\} \, \text{Re}(a_n b^*_n)] \, . \tag{4.12}$$

The value g is a crude measure of the degree of forward scattering and takes on a major role in the diffusion approximation. When g is small the scattering is fairly isotropic; when the phase function is strongly forward throwing g is closer to unity (§4.4.3). Some illustrations of the dependence of g on x are given in Fig. 4.4.

Finally we take note of the Henyey-Greenstein (analytic) phase function

$$P_{HG} = (1-g^2) \, (1+g^2-2g \cos \theta)^{-3/2} \tag{4.13}$$

which was devised to study the effects of changing g on various scattering problems. Note that for this function $p_n = g^n$. It has served a useful purpose in many such studies, but of course does not represent the real phase function accurately.

4.4. APPROXIMATION FORMULAE

4.4.1. The long-wavelength limit
As in the development of the approximation for the complex cross-sections outlined in §2.3.1 there are two approaches. The first involves viewing the scattered radiation as that radiated by the dipoles induced along three co-ordinate axes. Since the total amount of energy radiated by a dipole is well known, we can write immediately

$$C_{sj} = (8\pi k^4/3) \sum_{i=1}^3 |\alpha_{ij}|^2 \, . \tag{4.14}$$

This explains the λ^{-4} dependence of the extinction cross-section when m is real, since then scattering is the sole contributor. If m is not real then absorption will dominate; the albedo is very small (Fig. 4.4).

To evaluate the α_{ij}'s we recall the terminology adopted for the axially symmetric particle in §2.3.1, where α_{11} and α_{12} were given in terms of

α_A and α_B (eqn (2.33)). The only other dipole component induced involves

$$\alpha_{32} = (\alpha_B - \alpha_A) \cos \zeta \sin \zeta . \tag{4.15}$$

Therefore if we let

$$C_{sZ} = (8\pi k^4/3) \, |\alpha_Z|^2 \tag{4.16}$$

for Z=A, B it follows that

$$C_{s1} = C_{sA} , \quad C_{s2} = C_{sB} \sin^2\zeta + C_{sA} \cos^2 \zeta \tag{4.17}$$

as in eqn (2.34). Some expressions for Q_{sZ} appear in Table 2.1. Note that (except for infinite cylinders) $Q_{sA}/Q_{sB} = Q_{eA}/Q_{eB}$ when the imaginary part of the refractive index (n′) is small.

The alternative approach, through power series expressions for the expansion coefficients and substitution in equations like (4.9) for C_s, is also the only valid method for infinite cylinders. For spheres the first term in the expansion of C_s is that given in Table 2.1. It should be noted, however, that the real part of that term in C which varies as λ^{-4} is not identical to C_s unless m is real. In the case of infinite cylinders the values of C_s differ from those for short thin cylinders (see Table 2.1). Interference between waves scattered by adjacent lengths of the cylinder is responsible for the difference. For the same reason eqn (4.17) no longer holds, but it can be replaced by another simple relation if ever required.

The phase functions for spheres and spheroids in this limit will now be discussed. The underlying approach again is to examine the radiation pattern of the induced dipole(s). The angular distribution of the scattered field varies simply as the sine of the angle measured from the dipole axis. For spheres, using the notation of Fig. 4.2, we have

$$T_{11} = ik^3 \alpha, \quad T_{22} = ik^3\alpha \cos \theta \tag{4.18}$$

where α is the isotropic polarizability of the sphere. From this is derived what is often called the Rayleigh phase function (for natural light and isotropic particles):

$$p(\cos\theta) = 3(1 + \cos^2\theta)/4 = 1 + \tfrac{1}{2} P_2(\cos\theta) . \tag{4.19}$$

Notice that although g=0 the scattering is not isotropic.

For axially symmetric particles like spheroids we use the notation described in Fig. 4.1 and find

$$\underline{T} = ik^3 \begin{pmatrix} -\alpha_A\cos\phi & -\alpha_A\cos\zeta\sin\phi \\ \alpha_A\sin\phi\cos\theta & -\alpha_A\cos\zeta\cos\phi\cos\theta -\alpha_B\sin\zeta\sin\theta \end{pmatrix} . \quad (4.20)$$

The matrix $\underline{F}$ (including the phase function) can be constructed from this.

4.4.2. X-ray haloes

Using the anomalous diffraction formulae for Q_e and Q_a, as in eqn (2.36), the corresponding Q_s can be obtained. This approximation is useful for examining the trends of the dependence of Q_s for normal particles on the ρ parameter and on the amount of absorption in the particle in the same way as described for Q_e in §2.3.2. Recall that because x is assumed to be very large the angular distribution of the scattered radiation resembles a diffraction pattern and therefore is generally of little interest for real grains at optical wavelengths.

On the other hand, the limit in which $\rho \to 0$, corresponding to the limit of Rayleigh–Gans scattering in which x>>1, is applicable to the scattering of X-rays by interstellar grains and so will be mentioned here. The (small) efficiency factor for scattering has already been described (e.g. eqn (2.38)). In anomalous diffraction, evaluation of the scattering amplitude is based on Babinet's principle followed by application of the Huygens–Fresnel principle; the resulting integral over the grain profile in the limit $\rho \to 0$ describes a diffraction phenomenon for a non–uniformly illuminated aperture. In the appropriate limit of Rayleigh–Gans scattering the integration of the effects of the individual volume elements leads to the same formula. There are no polarization–related effects, so that one scattering amplitude suffices.

As in similar diffraction phenomena for large objects the radiation is concentrated in a narrow cone about the forward direction and the cone is least extended in the direction in which the particle is largest. These points are illustrated by the simple cases summarized in Table 4.1. There the amplitudes E, F, and G are, respectively, the Bessel functions of order

TABLE 4.1

SCATTERING AMPLITUDES FOR ANOMALOUS DIFFRACTION IN THE LIMIT $\rho \to 0$

Shape	Illumination	$S(\theta,\phi)$†
sphere	—	$G(x\theta)$
circular disk	along axis	$F(x\theta)$
circular cylinder	perpendicular to axis	$F(x\theta\sin\phi)\,E(y\theta\cos\phi)$

†θ is the scattering angle, and ϕ is an azimuthal angle measured from the cylinder axis.

1/2, 1, and 3/2, each divided by the first term in its series expansion.
The lower-order functions fall off somewhat more rapidly from the maximum 1
at argument 0, the half-intensity points being 1.4, 1.6, and 1.8 respectively.

The specific application to an X-ray halo around a cosmic X-ray
source can now be described. Basically, grains directly in front of the
source scatter X-rays out of the line of sight, whereas grains slightly
displaced from the line of sight scatter radiation into a second cone whose
apex is at the observer. The opening angle of this cone is also small
because of small-angle scattering, and so the scattered radiation would
appear as a halo around the source.

If the density distribution of grains is uniform around the line of
sight, then in the optically thin case the angular intensity distribution
in the halo has the same form as that from a single grain, except that it
is scaled down by a factor r, the ratio of the distance from the source to
the grains to the total distance to the observer. For amplitude functions
like those in Table 4.1 the angle α_c, measured from the centre to the
half-intensity point, is

$$\alpha_c \simeq (10^{-5} \text{ cm/a}) \ \lambda(\text{Å}) \text{ r arc min .} \tag{4.21}$$

About half of the energy in the halo is contained within the angular radius
α_c. The shape of the halo may be asymmetrical if the lateral density distribution of the dust is non-uniform ('cloudy') or if the grains themselves
are non-spherical and aligned.

Calculation of the total intensity of the halo is straightforward.
An X-ray point source initially of unit intensity is reduced to $\exp(-\tau_s -\tau_a$
$-\tau_g)$, where τ_g represents absorption by the interstellar gas. The halo
intensity arising from a uniform distribution is $1 - \exp(-\tau_s)$, further
reduced by $\exp(-\tau_a -\tau_g)$. The observed intensity ratio is therefore
$\exp(\tau_s)-1$. The appropriate Q_s is computed using the anomalous diffraction
approximation, or where possible the limit of the Rayleigh-Gans approximation, as described in §3.1.3. Consider as an example icy grains, with a
0.2 μ, for which $Q_s \simeq 0.2$ near 23 Å, the K edge of oxygen. For a source at
1 kpc τ_s would be typically 0.2, and so the total halo intensity would
amount to about 20 per cent of that of the apparent point source. The
characteristic radius α_c would be about 6 arc min.

Conversely, observations of X-ray haloes are potentially valuable for
investigating the properties of the grains. Measurement of the halo size
and shape would give an estimate of the grain size and degree of alignment.
Furthermore, the anomalous behaviour of τ_s, and hence of the relative halo
intensity, near the K edge of an atom in the grain (e.g. Fig. 3.4) would

provide a means of identifying the major constituents of the grain material. The technical requirements, for a panoramic detector with good energy
resolution placed at the focus of an X-ray imaging telescope in space, seem
at first sight quite demanding, but nevertheless they should be realized as
X-ray astronomy progresses.

4.4.3. Asymptotic values for large spheres

It is clear from many of the preceding figures that Q_e, Q_s, etc. tend
towards limiting values as x increases. Quite generally $Q_e = Q_s + Q_a = 2$.
We have also seen in the anomalous diffraction approximation (§2.3.2) that
$Q_a = 1$ for absorbing spheres ($\gamma \neq 0$). However, for m not close to 1, $Q_a < 1$,
because there is always a certain amount of reflection. For the purposes
of describing scattering phenomena of large spheres, what we have been
calling scattered radiation can be separated into two components: diffracted radiation, which is always strongly peaked in the forward direction
and contributes 1 to Q_s, and reflected plus refracted radiation which need
not necessarily be forward peaking. Let $\tilde{\omega}'$ and g$'$ be the albedo and asymmetry factor for the reflected plus refracted component. Clearly $\tilde{\omega}' = 1$
when $\gamma = 0$; however, when $\gamma \neq 0$ the refracted rays are attenuated, and only
reflection contributes to $\tilde{\omega}'$. Table 4.2 provides a summary of the interrelationships of $\tilde{\omega}'$ and g$'$, $\tilde{\omega}$ and g for the total scattered radiation, and
several other parameters. (The radiation pressure efficiency factor Q_r is
discussed in §8.3.) Three additional columns for special cases are
included on the right. Of general interest are the points that large
absorbing spheres are not completely black except when $x|m-1| \to 0$ and that
g$' \neq 1$ except in the limit of no reflection ($|m-1| \to 0$).

Numerical integration of the expressions for $\tilde{\omega}'$ and $\tilde{\omega}'g'$ that can be
derived using geometrical optics is straightforward. For large dielectric
spheres the dependence of g and Q_r on m is shown in Fig. 4.6. Only values
of $\tilde{\omega}$ and g are displayed for absorbing spheres since Q_r is so close to
unity.

TABLE 4.2
ASYMPTOTIC VALUES FOR LARGE SPHERES

Parameter	Diffracted Component	Reflected plus Refracted Component	Total	Dielectric Sphere	Anomalous Diffraction ($\tilde{\omega}'= 1$)	($\tilde{\omega}'= 0$)
Q_e	1	1	2	2	2	2
Q_s	1	$\tilde{\omega}'$	$1+\tilde{\omega}'$	2	2	1
Q_a	0	$1-\tilde{\omega}'$	$1-\tilde{\omega}'$	0	0	1
$\tilde{\omega}$	1	$\tilde{\omega}'$	$\frac{1}{2}(1+\tilde{\omega}')$	1	1	$\frac{1}{2}$
g	1	g$'$	$(1+\tilde{\omega}'g')/(1+\tilde{\omega}')$	$\frac{1}{2}(1+g')$	1	1
Q_r	0	$1-\tilde{\omega}'g'$	$1-\tilde{\omega}'g'$	$1-g'$	0	1

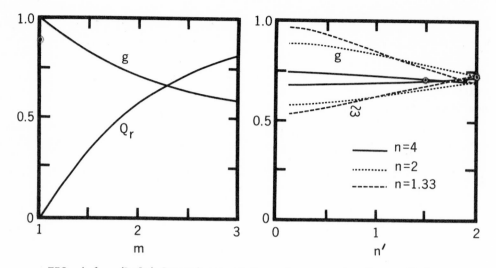

FIG. 4.6. (Left) Dependence of Q_r and g on m for large dielec-
tric spheres ($\tilde{\omega}=1$). (Right) Dependence of g and $\tilde{\omega}$ on m for
absorbing spheres ($Q_r \approx 1$).

4.5. POLARIZATION OF SCATTERED LIGHT

The transformation matrix $\underline{\underline{F}}$ for the Stokes parameters (§4.1) provides all
the information necessary for a complete description of the polarization of
the scattered light. Here a few of the important matrix elements are discussed in more detail.

   Polarization of natural light (spheres). Of major importance is the
production of polarization in a single scattering of natural light. The
elements F_{21}, F_{31}, and F_{41} give the amount of linear polarization (Q,U) and
circular polarization (V) respectively. Generally these elements are all
non-zero. However, for spherical particles there are no cross-polarization
scattering amplitudes and so $\underline{\underline{F}}$ simplifies to

$$\underline{\underline{F}} = \begin{pmatrix} \tfrac{1}{2}(M_2+M_1) & \tfrac{1}{2}(M_2-M_1) & 0 & 0 \\ \tfrac{1}{2}(M_2-M_1) & \tfrac{1}{2}(M_2+M_1) & 0 & 0 \\ 0 & 0 & S_{21} & D_{21} \\ 0 & 0 & -D_{21} & S_{21} \end{pmatrix}. \qquad (4.22)$$

Linear polarization therefore arises solely from the F_{21} component, and has
electric vector orientation parallel (positive Q) or perpendicular (negative Q) to the scattering plane (see Fig. 4.2). In an astronomical context
the electric vector is either along or perpendicular to the radius vector
joining the illuminating source and the patch of nebulosity on the sky.

The expression for the degree of polarization is

$$Q/I = F_{21}/F_{11} = (M_2 - M_1)/(M_2 + M_1) \tag{4.23}$$

which shows clearly the relationship to the scattered intensities of the
two modes.

In Fig. 4.5 curves of Q/I are given as a function of scattering angle
for the same size, wavelength, and refractive index combinations used for
the phase functions above. When x is small, the Rayleigh-scattering
approximation gives

$$Q/I = (\cos^2\theta - 1)/(\cos^2\theta + 1) \tag{4.24}$$

which is always negative. (Because of this interesting case the polarization is often defined to be $P=-Q/I$, a notation not followed here.) As x
increases, Q/I becomes positive over significant ranges in scattering angle
and is in general smaller. The single-particle resonances and ripples can
be dominant and distracting. Absorption within the particle again has
important effects.

In model calculations it is often desirable or necessary to examine
the combined effect of many particles distributed in size, orientation, and
space (scattering angle). Since Stokes parameters are additive, the ensemble average of the transformation matrix $\underline{F}$ is taken. One important application of this result is to a group of non-spherical particles oriented at
random. The averaged $\underline{F}$ has the same form as that for a single sphere (eqn
(4.22) and so the qualitative features of the above discussion apply.
Another application is to a size distribution of particles, which has the
effect of washing out the single-particle resonances seen in Fig. 4.5.

Some measurements of the phase function and polarization for roughened spheres (and also for cubic and octahedral particles) have been made
using the microwave analogue method.

Linear-to-circular conversion (spheres). Since $F_{31} = F_{41} = 0$ there
are no components U or V from natural light. However, the D_{21} term does
provide an important linear-to-circular (and circular-to-linear) conversion. Thus if the incident radiation is intrinsically polarized or has
become polarized in a previous scattering (a multiple-scattering process)
then circular polarization can be produced. Note that it is only through
the U component (at $\pm 45°$ to the scattering plane) that the phase differences of the two scattering modes can have an effect. Recall also that Q
and U are defined with respect to the scattering plane; in multiple-scattering problems the scattering plane is redefined for each scattering, and

so Q and U have to be transformed accordingly. (I and V are invariant under rotation of the co-ordinate system.)

Also shown in Fig. 4.5 is the quantity F_{43}/F_{11}, which gives the degree of circular polarization when I=U for the incident light. Small particles are inefficient at linear-to-circular conversion, but the efficiency for larger particles is much greater. Since the resulting V maintains the same sign over significant ranges in scattering angle and size, this process is of potential astrophysical importance.

Direct production of circular polarization (spheroids). If we relax the high degree of symmetry and look instead at spheroids we find that the transformation matrix is full, except for special orientations of the particle with respect to the incident and scattering directions. Although the direct production of linear polarization is of interest, especially since the combination of the F_{21} and F_{31} components can result in an electric vector at any position angle relative to the radius vector, we focus our attention on circular polarization. Again linear-to-circular conversion is important. However, most significant is the direct production of circular polarization in a single scattering of natural light, because F_{41} is nonzero. The astrophysical requirement is that the ensemble of grains being considered show a preferred alignment, since random orientation would average the effect out.

An example of the orientation dependence of F_{41} is provided by the analytic expression in the long-wavelength limit. From eqn (4.20), using the same notation,

$$F_{41} = k^4 \, Im(\alpha_A^* \alpha_B) \cos \zeta \sin \zeta \sin \phi \sin \theta . \qquad (4.25)$$

Thus $F_{41}=0$ if the angle of incidence is $0°$ (nose-on) or $90°$ (edge-on), if $\phi=0°$ or $180°$ (the two reference planes are the same), or if $\theta=0°$ or $180°$ (scattering direction along the symmetry axis of the grain). Note also that in this limit $F_{41}=0$ unless there is absorption in the particle material.

For larger values of x the directions $\zeta=0°$ and ϕ or $\theta=0°$ or $180°$ still result in $F_{41}=0$. However, even spheroids composed of pure dielectric material can now have F_{41} non-zero, since phase differences between different parts of larger particles are important.

The quantity F_{41}/F_{11} is clearly a measure of V/I of the scattered light. Its variation with θ on planes of constant ϕ (see Fig. 4.1) is shown in Fig. 4.7 for $\zeta=45°$. Also shown are $4\pi F_{11}/C_s$, F_{21}/F_{11} and F_{31}/F_{11}. These curves give some impression of how much circular relative to linear polarization might be expected if the degree of grain alignment in an

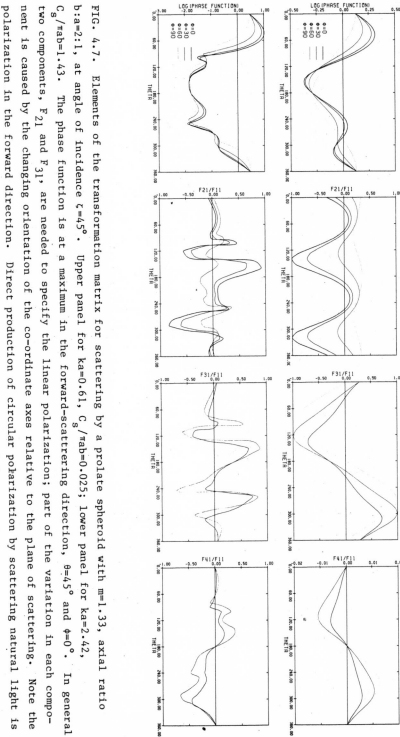

FIG. 4.7. Elements of the transformation matrix for scattering by a prolate spheroid with m=1.33, axial ratio b:a=2:1, at angle of incidence ζ=45°. Upper panel for ka=0.61, $C_s/\pi ab$=0.025; lower panel for ka=2.42, $C_s/\pi ab$=1.43. The phase function is at a maximum in the forward-scattering direction, θ=45° and φ=0°. In general two components, F_{21} and F_{31}, are needed to specify the linear polarization; part of the variation in each component is caused by the changing orientation of the co-ordinate axes relative to the plane of scattering. Note the polarization in the forward direction. Direct production of circular polarization by scattering natural light is demonstrated in the curves on the right; there is a large increase with ka.

ensemble is high. It is quite remarkable. Recall, however, that any diso-
rientation will reduce the circular component much more quickly than the
linear component.

FURTHER READING

§4.1. See §2.2.1 (van de Hulst 1957).
§4.2. See §2.2.
§4.3. See §2.2.1 (especially Wickramasinghe 1973).
§4.4.1. See §2.3.1.

§4.4.2
MARTIN, P.G. (1970). On the interaction of cosmic X-rays with interstellar
        grains. Mon. Not. R. astron. Soc. 149, 221.
ALCOCK, C., and HATCHETT, S. (1978). The effects of small-angle scattering
        on a pulse of radiation with an application of X-ray bursts and
        interstellar dust. Astrophys. J. 222, 456.

§4.4.3. See §2.2.1 (Kerker 1969).
§4.5. See §2.2.1.

5

REFLECTION NEBULAE

5.1. TYPES OF REFLECTION NEBULOSITY

A reflection nebula, as its name implies, owes its diffuse brightness to the scattering of incident light by the constituent dust particles. Such a reflection phenomenon occurs in a variety of distinguishable ways; we consider a division into three main types. The first and perhaps the most familiar type is the illumination of the surrounding interstellar medium by an intrinsically luminous star, as in the nebulosity in the Pleiades cluster (Fig. 5.1). This provides a means of studying the small-scale structure of the medium, which by the appearance of the nebulae is patchy and often filamentary. We shall refer to this type as 'interstellar' reflec-

FIG. 5.1. Reflection nebulosity in the Pleiades cluster (M 45). Of particular interest are the long parallel rays running east-west (N up and E left), and the curved striations around Merope (23 Tau, centre S) and Maia (20 Tau, centre N). Scale: 10 arc min = ————————. (Kitt Peak National Observatory 4-m photograph.)

tion nebulae. In the second type the dust is more intimately related to
the illuminating star, and the nebulosity is characteristically 'compact'.
Members of the third type are found at high galactic latitudes and appear
to be illuminated by the integrated light of the Galactic plane.

Interstellar reflection nebulae. Many nebulae discovered on sky survey photographs have been catalogued. They are typically one or more arc
minutes in extent with optical surface brightness in the range 20–22 mag
(arc sec)$^{-2}$. The detection limit is related to the sensitivity of the survey material. For example, the values of the limiting surface brightness
for the Palomar Sky Survey blue and red series are 23.6 and 22.4 mag (arc
sec)$^{-2}$ respectively. Amongst the observed nebulae there is an empirical
relation (found by Hubble) between the apparent brightness m_* of the star
and the angular extent a of the nebula:

$$m_* = -5 \log a + \text{constant} . \qquad (5.1)$$

In absolute terms, the radius of a nebula on the Palomar Sky Survey blue
plates obeys $\log R_B(pc) \simeq -0.2 M_V - 0.7$. The proportionality is that produced in the simplest of models using the inverse square law of illumination; apparently many nebulae are 'illumination bounded'. Theoretical
prediction of the constants in these relationships or the actual surface
brightness is of course more complicated, since it involves knowledge of
the dust distribution, the optical depth, the scattering phase function,
and the albedo.

Several lines of evidence can be presented to support the view that
these nebulae are representative of widespread structure in the interstellar medium as opposed to being some peculiarity associated with the illuminating stars. In the photographs on which reflection nebulae are found
there are often many dark patches of obscuration which no doubt would have
appeared as reflection nebulae had they been close enough to a bright star.
Next consider why it is that the illuminating stars are generally of spectral type B. The B stars are intrinsically luminous and so produce prominent nebulae. They are also sufficiently numerous that many nebulae can be
produced in a patchy interstellar medium simply by the chance proximity of
a star and a cloud of gas and dust; the fact that not all B stars have surrounding reflection nebulae (even in the Pleiades) lends support to this
picture. Main sequence stars of spectral type later than B are too faint
to produce a noticeable reflection nebula. However, there are some nebulae
illuminated by late-type giants and supergiants. Evidently it is not
necessary to identify the dusty material as debris left over from the formation of the illuminating star, although of course there are some cases in

which this would be the correct explanation.

A stronger correlation with the star-forming material might be expected for O stars, which are generally younger than B stars. However, O stars are also much hotter, so that, rather than create reflection nebulae from the clouds of material, they produce H II regions, in which emission lines dominate the nebular light. Reflection off internal dust in H II regions does produce a detectable polarized diffuse continuum, however. The presence of dust is also betrayed by internal extinction and thermal emission in the infrared (§9.4). The dusty gas clouds out of which such different optical nebulae are created, dominated by reflection or by emission, could otherwise be quite similar.

Interstellar dust is a good tracer of galactic spiral structure (§12.2.2). In our own Galaxy the concentration of both dust and B stars in the spiral arms produces associations of reflection nebulae (R-associations). Consequently the discovery of R-associations and subsequent measurement of their distances (by spectroscopic parallax of the illuminating stars) is a valuable procedure for mapping out spiral structure within a few kiloparsecs of the sun.

Compact reflection nebulae. Other cases of reflection nebulosity, usually compact, may be distinguished as being due to dust more closely related to the illuminating star. Those called cometary nebulae comprise a small group with peculiar morphologies, in which the star appears in a geometrically striking position. Two examples are Hubble's variable nebula (NGC 2261) which is conical, and the 'Red Rectangle' (CRL 915) which is biconical (bipolar); they are illuminated by the early-type emission stars R Mon and HD 44179 respectively. A compelling picture of such nebulae appeals to aspect-dependent extinction in a flattened dust cloud to explain the peculiar geometry (Fig. 5.2). In these compact nebulae much of the obscuring dust near the star is hot enough to be easily detected by its thermal emission in the infrared (§7.2.1).

Other smaller double cometary nebulae such as 'Minkowski's Footprint' (M 1-92) and the 'Egg Nebula' (CRL 2688, Fig. 5.3) might be explained by an edge-on geometry, similar to that proposed for the classical 'Egg-timer Nebula' (Lk Hα-208) but with the modification that the illuminating star is more completely obscured in the optical region. A schematic diagram of a possible configuration is shown in Fig. 5.2. Infrared measurements locate a strong thermal source in the dark central lane, as expected. Since the appearance of a symmetrical reflection nebula requires a special edge-on viewing aspect, there must be many more undetectable but physically similar sources.

The evolutionary status of these stars is uncertain,. though it has

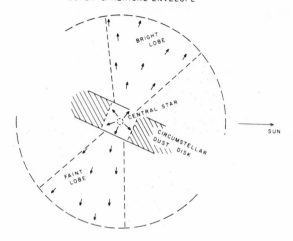

FIG. 5.2. Schematic diagram of the model proposed for M 1-92.
A central star exhibiting high-velocity mass ejection is sur-
rounded by an inclined disk of dust. A roughly spherical
envelope of lower-density dust is brightly illuminated by the
star in the two polar directions, possibly through actual holes
in the disk, as shown here. Extinction by the outer envelope
causes the polar fan farthest from the sun to appear the fainter. The star is seen obliquely through the disk and appears
just inside the inner edge of the bright lobe; it is very heavily obscured. The Doppler shifts of lines in the two lobes
indicate that material is streaming out through the lobes at a
velocity of about 30 km s^{-1} for the inclination of the disk to
the line of sight to the sun shown in this sketch. (From G.H.
Herbig, The spectrum and structure of "Minkowski's footprint":
M 1-92, Astrophys. J. 200, 1 (1975). University of Chicago
Press. $^{\circ}$1975 by the American Astronomical Society. All rights
reserved.)

been suggested that they represent various stages in the early evolution of
planetary nebulae. Related unresolved questions concern the relationship
of the dust to the star and whether the origin of the disk is common to
each object. Spectra of the reflected light, or of the central stars
directly where possible, reveal a wide variety of types, ranging from M5 I
(VY CMa) through F5 Ia (M 1-92) and A0 III (CRL 915) to O type (CRL 618).

There are many similarities in a geometrical interpretation that has
been suggested for a few semi-stellar patches of nebulosity with bright
low-excitation emission lines, which would otherwise be classified as Herbig-Haro objects. The illuminating object is completely obscured, indicat-

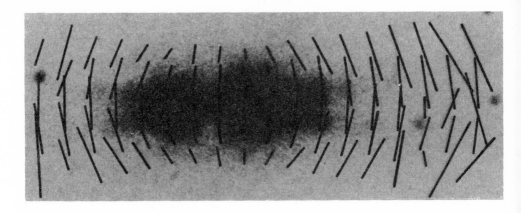

FIG. 5.3. Polarization map superimposed on a red-light (8000
Å) photograph of CRL 2688 (the Egg Nebula). The polarization
in the outer lobes is strong and oriented perpendicular to the
direction to the illuminating star. Scales: 10 arc sec =
────────; 50 per cent = ────. (From G.D. Schmidt,
J.R.P. Angel, and E.A. Beaver, Photoelectric polarization maps
of two bipolar reflection nebulae, Astrophys. J. 219, 477
(1978). University of Chicago Press. ©1978 by the American
Astronomical Society. All rights reserved.)

ing a large asymmetry in the dust distribution. Significantly offset from
the nebula is found a strong thermal infrared source, again originating
from hot dust. Identification of this as the illuminating source is confirmed using the direction of the polarization vectors of the nebular
light. The hidden emission-line objects, revealed only by reflection, are
thought to be very young, perhaps only a few hundred thousand years old.
They, and possibly some cometary nebulae, provide a unique glimpse of some
early stages of stellar evolution and of the state of the interstellar medium in the vicinity of star formation. We also take note of two presumably
young stars, FU Ori and V1057 Cyg (Lk Hα-190), which have infrared excesses
from hot dust. Following a dramatic brightening in the optical region each
now illuminates a small reflection nebula.

Excess infrared radiation shows that many late-type giants and supergiants have circumstellar dust clouds, thought to form during mass loss
from the stars (§7.2.1). These clouds do not generally have a large enough
extent (and often optical depth) to be detected as reflection nebulae. One
important exception is the small red nebula around VY CMa; whether the dust
is being produced by this star or is the remnant of a pre-stellar cloud is
not settled, though the former explanation is favoured in the context of
planetary nebula formation.

The study of the evolutionary relationship of the dust particles in
all types of compact nebulae to the particles in more diffuse interstellar
clouds is an important and active area of investigation at present.

High-latitude reflection nebulosity. Faint nebulosity, down to 25
mag (arc sec)$^{-2}$, has been discovered in several high-latitude fields. It
is sometimes diffuse and sometimes filamentary, and can often be traced
over regions more than $10°$ across. A consistent picture can be developed
in which clouds of dust (and gas) above the Galactic plane are reflecting
the integrated light of stars in the Galactic disk. (The height above the
plane might be about 100 pc.) Since the reflecting particles must also
cause extinction, deep photography offers a sensitive way of discovering
patchy extinction at high galactic latitudes, particularly near the polar
caps.

Keeping this wide variety of nebulae in mind, we shall now turn to
the observations and the scattering theory and models required to investigate the nature of the dust. Interstellar reflection nebulae can be used
to characterize further normal interstellar dust. In compact nebulae the
evolutionary changes associated with the different environments can be examined.

5.2. CONCEPTUAL MODELLING OF REFLECTION NEBULAE

Much of the difficulty in building a model of a reflection nebula lies not
in the scattering theory itself, but in the choice of the many parameters
on which the model depends. Consequently grids of models are often constructed to show the effects of varying each parameter. To facilitate numerical computations most models have treated only single scattering, although it is well known that for most reflection nebulae $\tau > 1$ in the optical
region. Nevertheless, many of the theoretical problems are illustrated by
the optically thin case.

Basic ingredients. Many of the factors are best considered with the
aid of a sketch such as Fig. 5.4. First there are some geometrical considerations, for which the distance to the nebula and its angular extent are
needed. Interstellar nebulae have linear sizes of a few parsecs in the
plane of the sky. The physical depth of the cloud and its placement
behind, beside, or in front of the star are unknown. They are very important parameters, however, since the geometry determines the range of the
scattering angle θ. The distribution of grains along the line of sight
through the nebula is also needed. In a slab model it might be taken as
uniform, or it might assume an R^{-2} dependence in a spherically symmetric

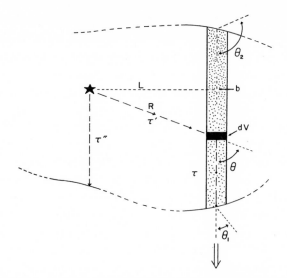

FIG. 5.4. Sketch of the plane of scattering for a particular
spot in a reflection nebula. Various parameters that have to
be chosen for the model are shown. (From B. Zellner, Dust
grains in reflection nebulae, in Interstellar dust and related
topics (eds. J.M. Greenberg and H.C. van de Hulst), Reidel,
Dordrecht (1973).)

model. The optical depths τ' and τ'' shown in the sketch are unimportant in
optically thin models. However, in more realistic models they would be
very important; therefore to model even a single spot in the nebula the
whole distribution of dust would have to be postulated. Although some
tilted-slab models have in fact taken into account the various τ's and the
accompanying reddening, the predictions are still not completely reliable
because multiple scattering was neglected.

Additionally the grain size and grain material must be decided. For
interstellar reflection nebulae logical choices are those corresponding to
grains which explain interstellar extinction and polarization. Unfortunately even these are not well known, but there are at least restrictions
on the size if the refractive index is chosen. The problem can also be
turned around; it can be hoped that despite all of the uncertainties comparison of models with observations will tell something of the nature of
the scattering particles. This is particularly important in the compact
reflection nebulae we have mentioned, where considerable differences in the
properties of the grains may occur.

A number of observable quantities can be computed from optically thin
models: the spatial dependence of the surface brightness (to within a
scaling factor if the optical depth is not actually specified), the wavel-

ength dependence of the surface brightness at any position (again to within
a scaling factor), the spatial dependence of the colour of the nebula, the
spatial dependence of the degree of polarization and its position angle,
and the wavelength dependence of polarization at individual positions.
These will be discussed in turn.

Surface brightness. Brightness depends not only on the total scattering optical depth (and hence the albedo) but also more subtly on the
phase function (F_{11} to be exact), because different scattering angles are
important in different volumes of the nebula. This is least important for
small grains because the phase function is quite isotropic without a pronounced forward peak (Fig. 4.5). (A recurrent theme will be the pivotal
role played by the grain size.) The brightness of course also falls off
with distance because of the inverse square law. Although measurement of
the absolute surface brightness is straightforward, it is seldom predicted
by the models. This is understandable for optically thin models in which
the optical depth is never specified, and in any case predictions are plagued by geometrical uncertainties.

Optically thin models generally produce steeper surface brightness
gradients than those observed. One example is the Merope Nebula model
which best reproduces polarization and colour data; the gradient predicted
is a factor of 2 or more too steep. However, it is known from observing a
star (HZ 1-371) through this nebula that $E_{B-V} = 0.36$, so that a multiple-
scattering model should be used. Monte Carlo scattering calculations of
more realistic cases have in fact been able to produce a flatter gradient
more closely matching the data. Unfortunately, the latter models did not
include predictions of colour and polarization properties, and so a completely consistent model has not been tested.

At a given spot in the nebula the surface brightness is a function of
wavelength. The brightness is usually referred to that of the star to
remove the spectral dependence of the stellar radiation. The remaining
wavelength dependence arises from the grains, again through F_{11}. The grain
size is quite critical. For small grains the phase function is relatively
independent of wavelength, and so F_{11} varies like the scattering cross-section, rising very steeply with decreasing wavelength. For larger particles
the total scattering cross-section rises less steeply to the blue; therefore smaller particles generally produce a bluer nebula than larger particles. The situation for larger particles is further complicated by the
wavelength dependence of the angular variation of F_{11} (Fig. 4.5). Naturally optical depth effects which occur in actual nebulae should also be
considered, since the optical depths are themselves quite wavelength depen-

dent.

   Colour. The behaviour of the colour of a nebula is closely related
to the above discussion. The colour is usually given relative to that of
the star to show the true effects of the grains. A nebula should generally
be bluer than the star as a result of selective scattering, unless the star
suffers less reddening than the nebula by some quirk of the geometry. (The
outer regions of the nebula might also be redder than the inner parts if
the starlight arriving there to be scattered is more reddened during transmission.) A spatial dependence of the nebular colours can be caused by
the wavelength dependence of the phase function, since different scattering
angles predominate at different offsets from the star. Such colour differences are more pronounced in (model) nebulae involving larger grains.
Again, models will not be entirely realistic without treatment of multiple
scattering.

   Polarization. Polarization calculations for single scattering by
spherical particles are straightforward. Here the scattering plane is
defined by the lines of sight to the star and to the position in the
nebula. Since $F_{31} = F_{41} = 0$, only the Q component of linear polarization
arises. For small particles the electric vector is always perpendicular to
the scattering plane because Q is negative; thus the electric vector is
perpendicular to the radius vector joining the patch of nebulosity and the
star. However, for sufficiently large grains F_{21} can change sign with
scattering angle (Fig. 4.5), and so sometimes the electric vector might lie
along the radius vector. Although by our convention Q would then be positive, this has often been called 'negative polarization'; both cases are
denoted 'radial', though the former would better be described as
'tangential'.
   Polarization vectors with a 'non-radial' orientation can arise if the
special symmetry is broken. This might arise in multiple scattering in an
asymmetrical geometrical configuration, or even in single scattering by
aligned non-spherical particles. In the former case wavelength dependence
of the opacity and phase function could produce a rotation of the position
angle with wavelength. In the latter case inspection of the signs of F_{21}
and F_{31} shows that the electric vector is rotated from its otherwise
'tangential' orientation towards lying along the long axis of the grain
profile, at least for normal-sized grains. This would be more or less perpendicular to the E vector of transmitted light. Circular polarization
might be expected in either case as well, by linear-to-circular conversion
(F_{43}) or by direct production (F_{41}) respectively.
   The polarization integrated over the range in scattering angles which

occurs along a line of sight through the nebula is obtained by averaging
F_{21} and F_{11} separately and then dividing. This takes proper account of the
common situation in which the scattering angles with the highest degree of
polarization do not contribute a major fraction of the intensity. Such
depolarization occurs even for small particles but becomes more pronounced
at larger sizes or shorter wavelengths, where F_{21} can have sign changes and
the phase function develops a sharp forward peak. Several important observational consequences might be expected. The spatial distribution of the
degree of polarization should show a minimum near the illuminating star
because a greater fraction of the light arises there from nearly forward or
backward scattering. Smaller grains should produce a smaller spatial gradient and an overall higher degree of polarization. Because the depolarization is generally greater at shorter wavelengths, the degree of polarization at any particular spot in the nebula should decrease to the blue; the
decrease would be slower for smaller grains. Multiple scattering introduces additional depolarization; a factor of 2 at $\tau \approx 1.5$ is not atypical.
All of these suggested trends might of course be violated in specially contrived geometries.

5.3. OPTICAL AND ULTRAVIOLET OBSERVATIONS

Large numbers of reflection nebulae have not been studied in detail in a
systematic way because of the difficulties inherent in measuring faint diffuse nebulosity. Considerable effort must be devoted to removing the
effects of sky brightness and starlight scattered in the telescope, and
even then the accuracy may be limited by lack of sufficient photons in a
reasonable integration time. Some examples of the types of observation
that have been carried out are given below, together with an indication of
how they might be interpreted.

Photography. There are optical photographs of all reflection nebulae, this being the principal mode of discovery. They provide fundamental
data on spatial extent, geometrical features, surface brightness, and sometimes polarization. Interstellar reflection nebulae are typically smaller
than a few parsecs, somewhat smaller than the length scales for cloud
structure inferred using other techniques (§9.2). They are also considerably more dense than the average interstellar medium. For instance if the
Merope Nebula is about as thick as its measured width the extinction coefficient is about 1 mag pc^{-1}, a factor of 10^3 larger than the average and
more than 10 times larger than in typical clouds. A consistent picture in
terms of a compressed cloud can be developed, however. Relatively dense
clouds are of course favoured in producing detectable reflection nebulae

since their surface brightnesses are higher. Even a cursory glance at many
of the pictures is enough to convey the irregular shapes of the interstellar clouds. Closer inspection often reveals filamentary structure, as in
the Pleiades nebulae (Fig. 5.1). The filaments there are quite thin, ranging in size from 0.5 down to 0.005 pc. The orientation of the filaments
changes across the nebula and apparently has no relation to the star.
Polarization in the filaments will be described below. It is possible that
in the Pleiades (and other reflection nebulae) we have, with the help of
illumination by a star, the opportunity of examining the structure of an
interstellar cloud in some detail. However, it has been suggested that
many of the finer features may instead be the combined result of radiative
and gravitational forces on the dust clouds during close encounters with
the stars.

The morphologically peculiar cometary nebulae have been classified
according to their conical, biconical, arc, or comma-like appearance as a
preliminary step in investigating their origin. A few examples have been
mentioned; a bipolar nebula is pictured in Fig. 5.3. The geometrical characteristics of some cometary nebulae argue for an interpretation in the
context of star formation, while others look more like the result of stellar mass loss, possibly related to the formation of planetary nebulae.
Both explanations have been put forward for some of the same nebulae!

Most photometry (and polarimetry) of other nebulae has been carried
out photoelectrically, spot by spot in the nebula. However, the optical
surface brightness of the Merope Nebula has been measured photographically;
the relatively slow decrease away from Merope seems to require multiple
scattering. With the advent of automated microdensitometers, photographic
plate material can now be used more routinely to make maps of the brightness and colours (or even the polarization) of much fainter nebulae (e.g.
NGC 1999 and NGC 2261). Further results from this technique are awaited
with interest.

Another nascent field is ultraviolet photography. An early photograph of the Orion constellation showed an unexpectedly high diffuse
brightness for the spectral range exposed, 2200–4900 Å. The huge nebulous
region, measuring about 19° by 14° and centred near the hot bright stars in
the Orion belt and sword, can be interpreted as reflection by interstellar
grains (as opposed to line emission). Comparison of the integrated diffuse
emission with the direct starlight indicates a high albedo, unless there
are other bright stars obscured from our point of view, as has been suggested elsewhere in a different context. The relatively flat spatial
dependence of brightness cannot be explained by a uniform distribution of
dust, but a thick ellipsoidal shell provides an adequate model. Much of
this shell would lie in the H I region outside the ionization front deli-

neated by Barnard's Loop, which appears most strongly as an arc of Hα emission on the eastern half of a similarly centred 14° by 10° ellipse. (In
the evolution of this extended H II region neutral dusty material is being
pushed out of the central parts ahead of the ionization front. Radiation
pressure on grains may contribute to the outward pressure (§8.3).) This
arc shows up in the ultraviolet nebulosity.

A subsequent photograph in the far ultraviolet (1230-2000 Å) reveals
a similar diffuse nebulosity in which the discrete Loop structure appears
with much less contrast. In a model based on a complete shell the
decreased prominence of the Loop might be attributed to an increase in g in
the far ultraviolet, so that scattering near 90° in the Loop is less favoured.

This explanation is not particularly unique. For example another
model, based mostly on photoelectric measurements of the spatial gradient
of the surface brightness in several ultraviolet passbands, attributes the
bulk of the nebulosity to backscatter off molecular clouds seen in this
direction. Both the wavelength dependence of the steepness of the brightness distribution and the changing contrast of the Loop structure are taken
as evidence for a decrease of g into the ultraviolet, continuing shortwards
of 1800 A. (A similar result for the Merope Nebula is discussed below.)
Further investigations will be required to ascertain the correct geometrical configuration before the grain properties can be derived conclusively.

Photoelectric photometry. More typical of photometric measurements
are photoelectric U, B, and V magnitudes as a function of position in the
nebula. Usually these are combined as a display of colour (relative to the
star's) versus offset, as shown in Fig. 5.5 for the Merope Nebula. Taking
into account the decrease in brightness with offset, the integrated nebular
light is bluer than the starlight (B6 IV), the result of selective scattering. The low reddening of Merope itself, as compared with other stars seen
through the nebula, indicates that the Merope nebula is illuminated from
the near side; this knowledge removes some of the uncertainty in the
modelling procedure. The reduced colour differences at larger offsets can
then be understood as the result of internal reddening and the predominance
of scattering angles nearer 90°. Note how the neutral colour difference
occurs closer to the star at shorter wavelengths, a trend that continues
into the near ultraviolet. It is more difficult to model the colour gradient of the nebula with extremely small particles because the wavelength
dependence of their phase function is insignificant. This raises the question of what size of grain might be expected. If normal interstellar
grains are involved then the size is dependent on the composition. As seen
in §3.1.2, dielectric icy grains, which have a lower refractive index, must

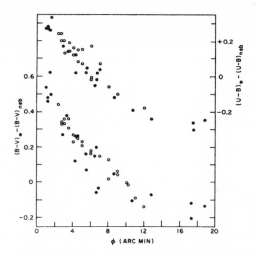

FIG. 5.5. Observed colour differences of the Merope Nebula
relative to Merope. The upper data are for U-B, and the lower
data are for B-V. The nebula becomes redder than the star at
large offsets. (From M.S. Hanner, Light from reflection nebulae. IV. Scattering by silicate grains, Astrophys. J. 164,
425 (1971). University of Chicago Press. °1971 by the University of Chicago. All rights reserved.)

have larger sizes than, for example, silicate grains to explain the interstellar extinction curve. Both sizes are significantly larger than for
'metallic' grains like graphite or iron, which would qualify as small
grains for the present discussion. This assumption relating composition to
size lies behind the conclusion that dielectrics provide a more satisfactory fit to the nebular colours (and polarization). Some caution must be
exercised before more exacting inferences are drawn from the data since the
models are only approximate.

A complementary photoelectric technique concentrates on accurate measurements of the wavelength dependence of surface brightness at a few
selected areas in the nebula. Among those studied is NGC 2068 (M 78), for
which data expressed relative to the stellar continuum (B1 V) are shown in
Fig. 5.6 (on left). (In this plot the zero point is arbitrary.) Again
the blue colour of the scattered radiation can be seen. A quantitive determination of model parameters from these data is not possible without sup-
plementary observations of, say, the wavelength dependence of polarization,
to which we shall return (see Fig. 5.6 (on right)).

A good illustration of the way in which information might be gained
from reflection nebulae is provided by the Merope Nebula. In the first
place the observational coverage is unusually complete. Existing ground-

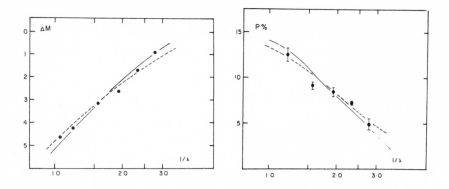

FIG. 5.6. Theoretical match to brightness and polarization
data for a region in NGC 2068 (ΔM is the magnitude of the
nebula minus the magnitude of the illuminating star, to within
a constant). The allowed range of scattering angles in the
model was 10-50°. Solid curves for m=1.5, average radius ∿ 0.06
μ; broken curve for m=1.1, average radius ∿ 0.12 μ. (From B.
Zellner, Dust grains in reflection nebulae, in <u>Interstellar</u>
<u>dust</u> <u>and</u> <u>related</u> <u>topics</u> (eds. J.M. Greenberg and H.C. van de
Hulst), Reidel, Dordrecht (1973).)

based measurements of the wavelength dependence and spatial distribution of
surface brightness have been supplemented by parallel observations out to
1550 Å using the ultraviolet photometer on ANS. To match this quality,
many multiple-scattering models for plane-parallel nebulae have been computed using a Monte Carlo code. For comparing theory with observations a
geometrical arrangement with 23 Tau in front of the nebula was adopted,
primarily on the basis of optical data together with the requirement that
$g≈\tilde{\omega}≈0.7$ there. The latter choices coincide with the results from diffuse
Galactic light and other scattering phenomena (§10.3.2). The wavelength
dependence of total optical depth was assumed to follow that of interstellar extinction, but g and $\tilde{\omega}$ were allowed to vary in the ultraviolet. It
turns out that nebulae involving largely backscatter do not provide a good
discriminant of $\tilde{\omega}$, but changes in g (along with τ) can produce significant
differences in the spatial brightness gradient at different wavelengths.
In particular the gradient steepens as g decreases and/or as τ increases in
these models (expressed in another way, these trends produce a reddening of
the nebular colour with increasing angular offset). A major conclusion of
this study of the Merope Nebula is that g decreases (from 0.7) into the
ultraviolet, particularly beyond 2200 Å. Although a value reaching as low
as zero is possible, the actual range of g cannot be determined precisely
without more detailed models.

The apparent decrease of g towards shorter wavelenths is probably not
the property of a single grain. It seems more likely that there is a distribution of grain sizes, and that the dominant size for scattering
decreases in the ultraviolet; this would allow a smaller effective x for
the scatterers and an accompanying smaller g (more isotropic scattering).

Polarimetry. Various polarization observations in the optical have
been made. Fig. 5.7, for NGC 7023, shows the typical pattern of
'tangential' polarization centred on the illuminating star. Recall that if
the grains are too large then the direction of the electric vector can
become 'radial', which has not been seen. Some models have suggested that
normal-sized silicate grains would be large enough to produce 'radial'
polarization in the blue. However, to be quite sure it would be desirable
to have a model with multiple scattering that matched both colour and
polarization data. This uncertainty in interpretation also applies to
other results to be discussed later.

As for several other nebulae, the data for NGC 7023 were obtained
photoelectrically for precision. A few nebulae, generally compact, are so
strongly polarized (50 per cent) that useful photographic measurements have
been made; the polarization maps of VY CMa are a good example. Subsequently photoelectric panoramic detectors have been used to produce polari-

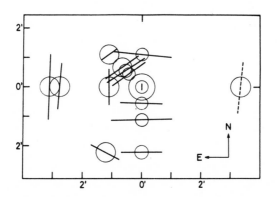

FIG. 5.7. Polarization of NGC 7023 showing large polarization
vectors oriented perpendicular to the direction to the illuminating star HD 200775. The polarization of the annulus surrounding the central star is small, and is represented by the
short bar at the centre of the annulus. Scale: (for blue
light) 10 per cent = ————. (From A. Elvius and J.S. Hall,
Color and polarization of reflection nebulae NGC 2068, NGC
7023, and Merope Nebula in three spectral regions, Lowell Obs.
Bull. (135) VI, No. 16 (1966).)

zation maps of similar highly polarized nebulae, like M 1-92 and CRL 2688
(Fig. 5.3). (In fact the polarization of the latter is so large that it
can be inspected visually by rotating a sheet of polaroid behind the eyep-
iece!) All of these maps have a 'tangential' polarization pattern that
points out the position of the illuminating star, which is quite a useful
property. In NGC 2068, where there are two bright candidates, the ambiguity is resolved. In bipolar cometary nebula like M 1-92, where the star is
visually obscured, the coincidence with a thermal infrared source in the
bisecting dust lane is confirmed. In the nebulous emission-line objects
mentioned above the presence of polarization not only establishes the
reflection nature of the nebulosity (polarization measurements in the light
of the emission lines would be valuable) but also adds to the understanding
of its geometry since an infrared source is found at the offset position
predicted by the polarization vectors.

The electric vectors in some filamentary interstellar nebulae, like
those in the Pleiades, deviate from the 'tangential' direction to lie perpendicular to the filamentary structure. Qualitatively this is the effect
expected if non-spherical grains are aligned so that the long direction of
the grain profile is perpendicular to the filament. No quantitative models
have been presented. However, empirical evidence for such an orientation
is provided in a few cases where the polarization of starlight transmitted
through the filament has been measured; the electric vector of the transmitted light is parallel to the filament. The mechanisms responsible for
this alignment and for the formation of the filaments are not well understood, though the interstellar magnetic field might be suspected to play an
important role. Dynamical effects should be considered too.

Photoelectric measurements of the degree of polarization as a function of wavelength or offset have often been made in conjunction with photometric observations to help restrict the model parameters as much as possible. An example of the former was given in Fig. 5.6 (on right), whereas
the latter is illustrated for the Merope Nebula in Fig. 5.8. Two trends
can be noted. First, the polarization decreases towards shorter wavelengths. As discussed in §5.2 this is difficult to explain using optically
thin models with small grains, but for larger grains or in optically thick
nebulae such depolarization is expected. Second, the polarization
increases with offset. Again, small grains would not produce as large an
effect as would larger grains. Quantitative results from optically thin
models suggest that although the grains are not small they are also not as
large as those inferred from the interstellar extinction curve. The origin
of this discrepancy might lie in the inadequacies of present models of
reflection nebulae.

Circular polarization has not yet been detected in interstellar

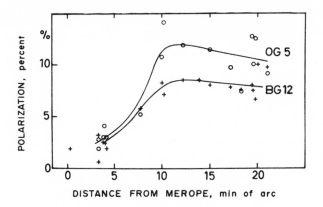

FIG. 5.8. Polarization of the Merope nebula as a function of
offset in two bandpasses (λ_{eff}^{-1} = 1.74 and 2.24 μ^{-1}). The
curves are guides only. (From A. Elvius and J.S. Hall, Color
and polarization of reflection nebulae NGC 2068, NGC 7023, and
Merope Nebula in three spectral regions, <u>Lowell</u> <u>Obs</u>. <u>Bull</u>.
(<u>135</u>) VI, No. 16 (1966).)

reflection nebulae. However, in the compact circumstellar region surrounding VY CMa the spatially integrated light shows significant circular polarization on the order of −0.1 per cent, reaching a maximum of −0.4 per cent
at 1 μ. Whether this wavelength dependence is the real effect of the
linear-to-circular conversion efficiency or is the result of depolarization
is not clear. Also since the nebula has not been spatially resolved in
circular polarization measurements the actual site of production is not
known. By analogy with even more compact infrared sources with similar
polarization, multiple scattering in the immediate circumstellar region can
be suspected.

CRL 2688 has a wavelength-independent circular polarization at optical wavelengths amounting to −0.6 per cent. Estimates based on the low
surface brightness and high degree of polarization both support a low scattering optical depth in the extended lobes, but multiple scattering in the
flattened central disk is not excluded. Explanations in terms of single
scattering off aligned particles or of birefringence from locally aligned
particles seem at least as contrived. The lack of variation with wavelength should ultimately be a useful diagnostic.

From this short exposition it will be clear that reflection nebulae
themselves are quite interesting phenomena. As for the nature of the
grains in the nebulae, more definitive statements should be possible when
more extensive measurements become available and if more sophisticated

models are developed. The anticipated rewards will not be reaped without a
major effort.

FURTHER READING

§5.1

JOHNSON, H.M. (1968). Diffuse nebulae. In Nebulae and interstellar matter
    (eds. B.M. Middlehurst and L.H. Aller). University of Chicago Press,
    Chicago.

HERBST, W. (1975). R associations I. U B V photometry and MK spectroscopy
    of stars in southern reflection nebulae. Astron. J. 80, 212.

STROM, S.E, STROM, K.M., and GRASDALEN, G.L. (1975). Young stellar objects
    and dark interstellar clouds. A. Rev. Astron. Astrophys. 13, 187.

CALVET, N. and COHEN, M. (1978). Studies of bipolar nebulae. V. The gen-
    eral phenomenon. Mon. Not. R. astron. Soc. 182, 687.

SANDAGE, A. (1976). High-latitude reflection nebulosities illuminated by
    the galactic plane. Astron. J. 81, 954.

§5.2

ROARK, T., ROARK, B., and COLLINS, G.W.,III (1974). Monte Carlo model of
    reflection nebulae: intensity gradients. Astrophys. J. 190, 67.

WITT, A.N. (1977). Multiple scattering in reflection nebulae. I. A Monte
    Carlo approach. Astrophys. J. Suppl. 35, 1.

§5.3

ARNY, T. (1977). A model for the filamentary structure in the Pleiades
    reflection nebulosity. Astrophys. J. 217, 83.

BRUCK, M.T. (1974). Photographic surface photometry of the nebulae sur-
    rounding V380 Ori and R Mon. Mon. Not. R. astron. Soc. 166, 123.

CARRUTHERS, G.R., and OPAL, C.B. (1977). Far-ultraviolet imagery of the
    Barnard Loop Nebula. Astrophys. J. Lett. 212, L27.

WITT, A.N. and LILLIE, C.F. (1978). Ultraviolet photometry from the Orbit-
    ing Astronomical Observatory. XXX. The Orion reflection nebulosity.
    Astrophys. J. 222, 909.

WITT, A.N. (1977). The Merope Nebula revisited: The phase function of
    dust grains in the ultraviolet. Publ. astron. Soc. Pac. 89, 750.

ZELLNER, B. (1974). Polarization studies of reflection nebulae. In Plan-
    ets, stars and nebulae studied with photopolarimetry (ed. T. Geh-
    rels). University of Arizona Press, Tucson.

6

ABSORPTION AND THERMAL EMISSION OF RADIATION

6.1. ABSORPTION CROSS-SECTIONS AND EMISSIVITY

Absorption cross-sections serve two basic purposes: they determine how much energy is removed from the radiation field at any given wavelength, and via Kirchhoff's law affect the spectrum of the thermal radiation emitted. These functions are often intimately linked because absorption of starlight is important in maintaining the grain temperature (§6.2). We have already seen that in many compact reflection nebulae dust close to the illuminating star is heated up sufficiently to become a prominent infrared source. Similar thermalization of optical and ultraviolet radiation is detectable in circumstellar envelopes (§7.2), in H II regions and planetary nebulae (§9.4), in the solar system (§9.5), and generally in the Galactic plane (§7.1). Absorption also influences the diffuse radiation field within a dense dust cloud (§9.3); its effects are usually expressed through the albedo.

Absorption cross-sections. Derivation of the absorption cross-section C_a has already been mentioned in several places above and so will only be reviewed here. The most common technique is to subtract C_s from C_e whenever these two cross-sections have been computed independently. Included in this category are the differential equation formulation for spheres, spheroids, and infinite cylinders, the approximation for large spheres (§4.4.3), and in principle the integral equation formulation, the discrete dipole array method, and the microwave analogue method (see §§2.2 and 4.2). In other cases C_a can be calculated directly; the discrete dipole array and anomalous diffraction (§2.3.2) approximations are two examples. The expressions for Q_e contained in Table 2.1 for the long-wavelength limit correspond in fact to Q_a; recall that in this limit Q_a dominates over Q_s unless m is real, in which case $Q_a = 0$.

Examples of the wavelength dependence of Q_a can be seen in Fig. 4.3.

Emissivity. The single-particle emissivity is

$$E(\lambda) = C_a(\lambda) \ B(T_d, \lambda) \qquad (6.1)$$

where $B(T_d, \lambda)$ is the Planck function and T_d is the grain temperature. The emitted spectrum is therefore a blackbody curve weighted by C_a. The characteristic peak can be shifted somewhat from that given by the Wien dis-

placement law, usually towards shorter wavelengths. It is also important
to note that any spectral features in C_a will be imprinted on the emissivity.

Let us be more specific. Dust particles are generally cool and so
emit most of their radiation in the infrared. Therefore the long-wavelength approximation can be used for normal-sized grains; for spheres (Table
2.1)

$$Q_a = -4x \; Im\{(m^2-1)/(m^2+2)\} \; . \tag{6.2}$$

In the infrared many materials not only have broad-band absorption, but
also have prominent absorption bands (§10.2). It can be seen from eqn
(6.2) that Q_a will generally increase when n′ does. (As a result of the
depolarizing field set up in the particle a strong feature can be slightly
displaced towards shorter wavelengths.) Hence the absorption bands which
characterize a material will show up as emission features and can be used
to identify the composition. A prerequisite is laboratory measurement of
m.

Non-spherical grains have absorption cross-sections that are dependent on the polarization and angle of incidence of the incoming radiation.
Correspondingly the emission by a non-spherical grain varies with direction
and can be polarized. A convenient way to treat this is with the cross-
sections C_{aI} and C_{aII} with respect to a reference frame like that illustrated in Fig. 2.1. The angle of incidence needed for the calculations is
simply the angle of emission. The degree of polarization would be

$$Q/I = -(C_{aI}-C_{aII})/(C_{aI}+C_{aII}) \; . \tag{6.3}$$

Usually $C_{aI} > C_{aII}$ in the infrared. For the same grain it is likely that
$C_{eI} > C_{eII}$ in the optical region (§3.2). Therefore the orientation of the
electric vector of the emitted radiation (negative Q) would be perpendicular to that of light transmitted through the same aligned grains (positive
Q).

One important feature to note is that in the long-wavelength limit
grains emit in proportion to their volume (cf. eqn (6.2) from which $C_a \propto$
a^3). Therefore the total emission in an optically thin source provides a
measure of the mass of grains present, regardless of the actual size distribution. This is complicated by the fact that any thermal radiation
detected will be the integrated emission from a large number of grains with
different temperatures. Since the temperature can be a function of a
grain's size and its position in space, the size and spatial distributions
must often be specified before proper interpretation of the spectrum is

possible.

6.2. TEMPERATURE OF THE DUST

Consider an interstellar grain in thermal equilibrium with its surroundings
so that it radiates just as much energy as it receives from all sources.
The total energy radiated is obtained by integrating the emissivity (eqn
(6.1)) over wavelength and solid angle:

$$E_e = \int_0^\infty 4\pi \bar{C}_a(\lambda) \; B(T_d, \lambda) \; d\lambda \qquad\qquad (6.4)$$

where $\bar{C}_a$ is the absorption cross-section averaged over orientation. This
is to be equated to the energy input from three distinct sources: radiation, gas impacts, and exothermic surface reactions. If the radiation
field in which the grain is situated is isotropic, with energy density
$u(\lambda)$, then the total radiation absorbed may be written

$$E_r = \int_0^\infty \bar{C}_a(\lambda) \; c \; u(\lambda) \; d\lambda \; . \qquad\qquad (6.5)$$

The dominant contribution in eqn (6.5) generally comes from much shorter
wavelengths than it does in eqn (6.4), and so $\bar{C}_a$ must be known over a very
wide range in wavelength.

In unobscured interstellar space $u(\lambda)$ can be crudely approximated as
a blackbody spectrum at temperature 10^4 K with dilution factor $W=10^{-14}$. An
improved approximation involving several diluted blackbodies could be fitted to the data for analytical calculations, or the actual (tabulated) data
could be used for more exact calculations. The total energy density at
optical and ultraviolet wavelengths is about 8×10^{-13} erg cm^{-3} (0.5 eV
cm^{-3}). A blackbody immersed in this radiation field would have a temperature of 3.2 K. An interstellar grain has a somewhat higher temperature
because at optical wavelengths, where absorption occurs, Q_a is relatively
constant and near unity, whereas at infrared wavelengths, where the grain
emits, $Q_a \ll 1$ because the grain size is so much smaller than the wavelength.

For the purposes of a numerical solution and later reference, let us
assume that on average $Q_a = 0.5$ in the optical and ultraviolet so that $E_r =$
$\sigma_d 1.2 \times 10^{-2}$ erg s^{-1}, where σ_d is the mean geometrical cross-section of
the grain. Furthermore let us assume that $Q_a = 0.1(\lambda/10 \; \mu)^{-2}$ in the far
infrared, so that $E_e = \sigma_d 1.8 \times 10^{-10} T_d^6$ erg s^{-1}, or $T_d = 20$ K. Inside a
dense interstellar dust cloud the ultraviolet and optical radiation is sev-
erely attenuated; the consequent drop in E_r is partly compensated by an
increased contribution from the thermal emission by nearby grains. Since
T_d varies as a small power of E_r the corresponding decrease in temperature

is not large; a reduction by a factor of 2 at $A_V=10$ and a factor of 5 at $A_V=10^4$ would be typical (Fig. 9.10). Furthermore a lower limit of $T_d \simeq 3$ K is set by the cosmic microwave background radiation. In a circumstellar region the dust can be much hotter (Fig. 7.4). Again radiative transfer effects are important. An upper limit to T_d is set by the temperature at which vaporization becomes important.

Additional heat sources. Although radiative heating is often dominant there are several interesting, if exceptional, situations in which alternative sources of heating should be considered. Impacts by gas atoms, ions, and molecules impart energy if the mean kinetic energy of the impacting particle is greater before than after the collision. This is the case if the gas kinetic temperature T_g exceeds T_d (generally true), and if the gas 'atoms' stick to the grain long enough to become thermally accommodated so that they evaporate at a speed characteristic of the grain temperature. The efficiency of the accommodation process, or the sticking coefficient S, is probably large (0.3-1; see §8.1). The energy deposition per grain is

$$E_c(in) = \sigma_d \epsilon \, S \, n_a \, v \, \tfrac{1}{2}m_a v^2 \qquad (6.6)$$

where n_a is the space number density of the 'atom', m_a is its mass, v is its velocity before the collision, and ϵ measures the modification to the geometrical cross-section necessary for encounters of charged grains with ions (eqn (8.10)). A similar expression may be written for energy loss as atoms evaporate. If the net E_c is evaluated for appropriate Maxwellian distributions, assuming the net particle flux is zero, then

$$E_c = \sigma_d \epsilon \, S \, n_a \, 4(2kT_g/\pi m_a)^{\frac{1}{2}} \, kT_g \, (1-T_d/T_g) \, . \qquad (6.7)$$

If we consider as a numerical example hydrogen at $T_g=10^4$ K and take $\epsilon S \approx 1$, then $E_c = \sigma_d n_a \, 4 \times 10^{-6}$ erg s^{-1}. This is usually lower than the free-space value of E_r but should be considered in dense H II regions. For interstellar H I regions, where $T_g \approx 10-10^2$ K, E_c is a negligible source of grain heating. Nevertheless this same process may be a dominant mode of cooling the gas, especially in dense molecular clouds where (CO) line radiation can become trapped. Another potential application is to hot (10^7 K) gas in clusters of galaxies, if grains can survive there. Then collisions would not only dominate the grain heating but could also be the principal mechanism by which the gas cools (§12.3).

Among the exothermic reactions which occur on grain surfaces are proton-electron recombination and hydrogen-molecule formation, in which 13.6 and 4.48 eV are released respectively. Some fraction f of this energy will

be transferred to the crystal lattice, thus heating the grain, while the
rest will appear as excitation energy and/or kinetic energy of the product
of the reaction. Of the potential reactants which stick to the surface a
fraction γ will actually be resident long enough for a reaction to take
place. Since the impact energy input is only about $10^{-4} T_g$ eV per 'atom',
these reactions will undoubtedly contribute more than E_c. In an H II
region recombination may lead to an order-of-magnitude increase over E_c, so
that the density at which non-radiative heating dominates could be as low
as $n_{H\ II} \approx 10^2 - 10^3$ cm^{-3}. (This process is negligible as far as recombination
is concerned.)

In H I regions formation of molecular hydrogen on grains contributes

$$E_m = \sigma_d \gamma f S\ n_H\ 4.7 \times 10^{-8}\ T_g^{\frac{1}{2}}\ \text{erg s}^{-1}. \tag{6.8}$$

On setting $S=0.3$, $\gamma=1$, $f=0.5$, and $T_g=50$ K, we find $E_m = \sigma_d n_H\ 5 \times 10^{-8}$ erg
s^{-1}, which is still much less than E_r for free space. However, in dense
clouds, where n_H is high and E_r is lowered by extinction, E_m may prevent
the grain temperature from sinking to 3 K. For example if $A_V \approx 10$, E_r may be
down by a factor 10^2, so that if $n_H \approx 3 \times 10^3$ cm^{-3} (the molecular density
would be larger) then $E_m \approx E_r$. In a denser, more opaque cloud E_m would probably dominate, thus hindering a further decline in temperature such as that
described for $A_V \approx 10^4$.

Temperature spikes. Our discussion of radiative balance leading to
an equilibrium grain temperature, T_e say, makes the implicit assumption
that the total heat energy of the grain greatly exceeds the energy added by
absorption of an optical or ultraviolet photon. At low temperatures for
sufficiently small grains this assumption ceases to be valid, as is illustrated in Fig. 6.1 for small spheres (a=0.005 μ) of various pure materials.
(Also included is the enthalpy-temperature curve for an ideal crystal,
which is useful for analytic approximations.) For these grains the specific heat is clearly so low that absorption of an energetic photon raises
the temperature substantially above T_e. The grain equilibrates at this new
temperature in a matter of a nanosecond and proceeds to cool at a rate corresponding to its instantaneous temperature. For these small grains it is
found that a typical cooling time is of the order of minutes, much shorter
than the time between successive absorptions which is more like an hour in
interstellar space. Consequently the grain temperature sinks well below
T_e. The spiky behaviour of T is shown schematically in Fig. 6.2.

Since each of the processess involved can be described mathematically
it is possible to determine the relative amounts of time a grain spends
along the curve, and thereby obtain the time average of any function of T.

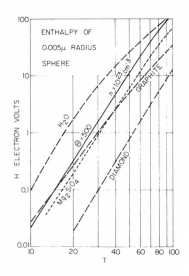

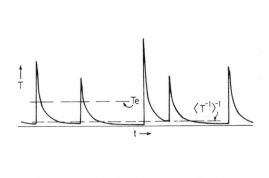

FIG. 6.1. The total heat energy in a small solid sphere of
radius 0.005 μ. (From E.M. Purcell, Temperature fluctuations
in very small interstellar grains, Astrophys. J. 206, 685
(1976). University of Chicago Press. ©1976 by the American
Astronomical Society. All rights reserved.)

FIG. 6.2 Schematic representation of the temporal behaviour of
the temperature of a small grain. T_e is the equilibrium temperature the grain would have if the energy were deposited con-
tinuously. The harmonic mean temperature is much lower. (From
E.M. Purcell, Temperature fluctuations in very small interstellar grains, Astrophys. J. 206, 685 (1976). University of Chicago Press. ©1976 by the American Astronomical Society. All
rights reserved.)

For example with the λ^{-2} dependence of Q_a described above, $T_e = \langle T^6 \rangle^{1/6}$ when
radiative cooling alone is considered. Evaporative cooling might be
included also if the evaporation rate were known (§ 8.1). Since the evaporation rate depends exponentially on T, the effective temperature is much
higher than T_e. On the other hand, the harmonic mean temperature is lower
than T_e, which is of possible significance to magnetic alignment of paramagnetic grains.

The spiky behaviour becomes suppressed rapidly with increasing grain
size for two reasons: the specific heat is much higher and so the height
of the spike is decreased, and the time between successive absorptions is
much reduced. In fact for a≈0.01 μ the time dependence of T should be
quite smooth. Furthermore, any imperfections in the structure of the

material can be expected to raise the specific heat markedly, so that thermal spikes may not even be important for very small interstellar grains.

FURTHER READING

§6.2
TABAK, R. (1977). The maximum temperature of interstellar grains. Astrophys. Space Sci. 46, 175.
DRAPATZ, S. and MICHEL, K.W. (1977). Optical and thermal properties of perturbed interstellar grains. Astron. Astrophys. 56, 353.

7

INTERSTELLAR AND CIRCUMSTELLAR THERMAL EMISSION

7.1. FAR-INFRARED EMISSION FROM THE GALACTIC PLANE

One frontier area being actively pioneered is far-infrared observation of
thermal emission from interstellar dust. On a Galactic scale this offers
the possibility of discovering the amount of dust, its temperature, and its
large-scale spatial distribution. The results of initial scans of the
Galactic plane in the 60-300 μ passband, made with a balloon-borne telescope, are shown in Fig. 7.1. Several points of interest are revealed.
Quite prominent are numerous localized sources with relatively high surface
brightness. These are identified with visible H II regions or closely
associated molecular clouds with embedded young stars, all regions which
contain copious amounts of warm dust (§9.4). Especially interesting is the
Galactic centre region; it consists of several such discrete components
superimposed on a diffuse background of radiation thought to originate in
cooler dust heated by late-type stars in the Galactic bulge. It is estimated that the combined infrared luminosity exceeds 2×10^9 $L_\odot$, representing substantial conversion of optical and ultraviolet stellar radiation.

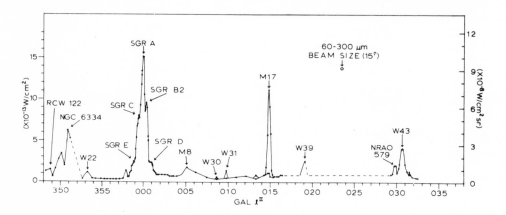

FIG. 7.1. Far-infrared scan in the Galactic plane revealing
discrete sources associated with H II regions and molecular
clouds, and possibly a small residual diffuse background from
widely distributed interstellar dust. (From F.J. Low, R.F.
Kurtz, W.M. Poteet, and T. Nishimura, Far-infrared scans of the
Galactic plane, Astrophys. J. Lett. 214, L115 (1977). University of Chicago Press.
)

Of most interest here is the residual far-infrared background in the Galactic plane, of order 3×10^{-9} W cm^{-2} sr^{-1} or less. To see whether this amount is consistent with radiation from widespread cool interstellar dust we must estimate the infrared optical depth and surface brightness expected. If the local value of the visual extinction coefficient, 2 mag kpc^{-1}, has wider applicability then $\tau_V \simeq 37$ over a pathlength of 20 kpc. Next τ_{IR} can be obtained by scaling with the ratio of the appropriate values of Q_e. Using the same illustrative values as above, $\tau_{IR} \simeq 0.02$ $(\lambda/100 \ \mu)^{-2}$. Thus the Galaxy is optically thin in the far-infrared spectral range, and the surface brightness is simply $\tau_{IR} B(T_d,\lambda)$. Integrating this, using τ_{IR} from our example, $I_{IR} \simeq 3 \times 10^{-17} T_d^6$ W cm^{-2} sr^{-1}. Hence consistency with the observations requires $T_d \simeq 21$ K, a not unreasonable value. (Adoption of a λ^{-1} dependence of Q_a still leads to the same temperature, $T_d \simeq 20$ K, if the 100-μ optical depth is chosen to be the same.) Recall that the spectrum will have the form $\lambda^{-n}B$ when $Q_a \propto \lambda^{-n}$, so that according to a generalized Wien's law the spectral peak occurs at

$$\lambda_p \simeq 0.3 \ T_d^{-1} \ 5/(n+5) \ \text{cm} . \qquad (7.1)$$

For $T_d \simeq 20$ K most of the energy therefore falls within the 60–300 μ passband and so the derivation is self-consistent.

Although this establishes the plausibility it is clear that quite a range of parameters could be accommodated, so that a more sophisticated treatment would be needed for either prediction or interpretation. The basic ingredients in any model are the efficiency factor Q_a, the grain temperature T_d, and the spatial distribution of the grains, all of which introduce uncertainty. In particular Q_a depends not only on the grain size but on the grain composition. Eventually the wavelength dependence of Q_a might be obtained from the long-wavelength spectrum, which would have the power-law form λ^{-n-4} for our simple example. The spectrum can, however, be distorted by contributions from grains with different temperatures. In any case laboratory studies of possible grain materials, including those with impurities and imperfect crystal structure, will be needed for a complete understanding.

A range in T_d will be introduced by both the presence of a variety of grains (through Q_a) and the differing environments in which the grains are found. Naturally grains near hot stars will be warmer and so will contribute predominantly in the shorter-wavelength portion of the spectrum. This distortion might be promoted by the correlation of dust and regions of star formation. On the other hand, shielding of grains in dense molecular clouds which seem to precede star formation will produce the opposite effect. Excitation temperatures of CO in molecular clouds indicate that T_d

$\geqslant 7$ K. (CO is excited by collisions with H_2; the kinetic temperature of H_2 is maintained by collisions with grains which are radiatively heated.) Thus the observed far-infrared spectrum should be broader than that from any single grain. Possibly the main heat sources for grains can be determined by examining the galactic latitude distribution of the thermal radiation.

Optical data are of limited use in mapping out the distribution of dust in the Galactic plane because of high obscuration. It might be supposed, however, that the dust can be traced out indirectly using maps of that ubiquitous molecule, CO. Certainly there is dust where CO is found, but the reverse need not hold. To investigate the usefulness of this hypothesis detailed measurements of the longitude and latitude dependence of the infrared surface brightness will be needed at several wavelengths to sort out the effects of varying temperature. If the correlations with the CO maps are good, then the relationships between the column densities of dust, CO, and H_2 (also H I) calibrated locally (§§9.2 and 9.3) would become useful. Location of the dust along the line of sight would follow from radial velocity information for CO, in the same way that spiral features are mapped out.

7.2. CIRCUMSTELLAR DUST

7.2.1. Stars with infrared excesses
Circumstellar dust causes extinction, in some cases so much that the underlying source becomes optically invisible. Hot dust close to the star also produces thermal infrared emission in excess of that arising from the stellar continuum. The presence of circumstellar dust is therefore indicated by an infrared excess, usually accompanied by reddening of the starlight. (Some early-type stars have infrared excesses that can be explained by free-free emission from a relatively dense circumstellar region; these can be distinguished by their spectra, and will not be discussed further.)

Circumstellar extinction. Circumstellar extinction can differ from normal interstellar extinction if, because the angular size of the dust distribution is smaller than the observing aperture, light scattered by the grains is included in the flux measurement. In the simple case of an optically thin spherically symmetric distribution that is smaller than the beam size, light removed from the direct radiation by scattering is exactly compensated by scattered radiation from the rest of the dust; hence the wavelength dependence of the extinction observed would follow the absorption cross-section rather than the extinction cross-section. Consequently the interstellar extinction law would not hold even if the grains were the same.

In optically thick envelopes proper radiative transfer solutions must be made, taking into account the phase function and the albedo, and there too the extinction law would not match the extinction cross-section. Further complications are introduced in asymmetric geometries, as seen above in cometary nebulae (e.g. Fig. 5.2). The optical light seen in such cases is a combination of reflected light from the envelope over the polar regions of the disk and some highly reddened radiation from the star, the relative contributions depending on the aspect.

Because of the problems arising from accurate treatment of the radiative transfer and uncertainties in geometry, circumstellar extinction has not been analysed in great detail. Generally there is a large amount of reddening. For stars without resolved reflection nebulosity this reddening must somehow be distinguished from the similar effect of interstellar reddening. From optical observations alone this is not easy, though in some particular cases, for instance for stars at high galactic latitude, a circumstellar origin for large amounts of reddening may seem most reasonable. Confirming evidence in the form of an infrared excess can be sought. This is illustrated by the systematic infrared (2-18 μ) study of the optically unidentified objects (about 30) in the 2-μ Catalog of IRC sources. Flux distributions for about 90 per cent of the sample were found to be like those of luminous late-type stars, lacking in infrared excess, and so the high reddening and optical extinction must be interstellar. The other 10 per cent showed infrared excesses, especially at longer wavelengths in the 10 and 18-μ bandpasses; subsequent study has demonstrated these to be cool luminous M and carbon (C) stars as well, with excessive amounts of circumstellar dust.

IRC and CRL sources. The detailed study of excess infrared emission provides valuable information on circumstellar dust. To find stars with such emission there have been basically two approaches. The first, searching through catalogues of infrared sources, has just been mentioned. In addition to the optically unidentifed objects described, many IRC sources are optically visible. Some which are classified as late M and C stars (giants and supergiants) have infrared excesses from envelopes ranging from optically thin to optically thick at 10 μ. Many of these stars, though not all, are Mira or semi-regular long period variables.

From measurements in the 10-μ region there is a spectroscopic distinction between dust surrounding an oxygen-rich as opposed to a carbon-rich star, consistent with the idea that the circumstellar dust was formed by the underlying star. Oxygen-rich stars show a 10-μ feature attributed to silicate material (see §10.2.2 concerning this and the 20-μ feature).

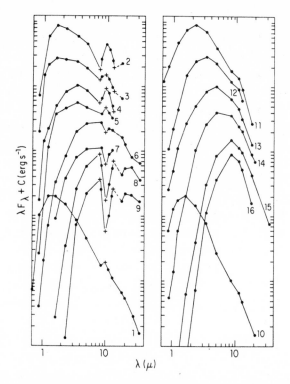

FIG. 7.2. Infrared spectra of oxygen-rich stars (on left) and
carbon-rich stars (on right) illustrating the range of spectral
behavior typically observed. The dots and pluses are broad-
and narrow-bandpass data respectively. Sources shown are as
follows: (1) o Cet, (2) IRC +10322, (3) NML TAU = IRC
+10050, (4) WX Ser = IRC +20281, (5) IRC +40004, (6) NML CYG
= IRC +40448, (7) CRL 490, (8) CRL 2591, (9) CRL 2205, (10)
UU Aur, (11) R Lep, (12) IRC +40485, (13) IRC +50096, (14)
IRC +40540, (15) IRC +10216, and (16) CRL 865. (From T.W.
Jones and K.M. Merrill, Model dust envelopes around late-type
stars, Astrophys. J. 209, 509 (1976). University of Chicago
Press. ©1976 by the American Astronomical Society. All rights
reserved.)

As shown in Fig. 7.2, this feature is in emission for shells that are optically thin at 10 μ (e.g. NML Tau = IRC +10050) but is in absorption if the
10-μ optical depth is large (e.g. NML Cyg = IRC +40448). In contrast,
shells around carbon-rich stars have a fairly smooth continuum at 10 μ and
a broad emission feature at 11.5 μ (§10.2.4). The underlying continuum and
the emission feature are thought to be due to graphite and SiC particles
respectively. IRC +10216 is a well-studied example. A range of increasing

optical depth is indicated by a progressive narrowing of the spectrum as
more of the near-infrared radiation from the star and hot (inner) dust is
blocked out (Fig. 7.2).

Atmospheric-band differences in the 2–4 μ region can also be
detected. M stars have 1.9 and 2.7-μ bands of H_2O, whereas C stars have a
characteristic, but unidentified, band at 3.07 μ. Spectrophotometric
results for some of these stars are shown in Figs. 10.3 and 10.5. Together
with 10-μ data these different near-infrared characteristics are sufficient
to classify other infrared sources for which no optical spectrum is available. For example a significant fraction of the sources in the AFGL/CRL
Infrared Sky Survey Catalog (survey at 4, 11, and 20 μ; CRL sources) appear
to be extreme examples of luminous late-type stars with thick circumstellar
shells.

Other CRL sources fall roughly into two classes. The first have
infrared spectra indicative of high obscuration, a deep 9.7-μ silicate feature, and substantial 3.1-μ absorption attributed to ice grains (Fig.
10.3). They are distinguished from the cool stars with self-absorbed silicate features by their lack of the near-infrared atmospheric features mentioned and by their association with molecular clouds. A good example is
the Becklin-Neugebauer object (BN) in the Orion Molecular Cloud. It seems
reasonable to identify the central sources as young objects. Cooler dust,
possibly within the molecular cloud, is believed to be responsible for the
3.1-μ and 9.7-μ features. This cold material, which contains ice, probably
bears a closer resemblance to normal interstellar dust than to circumstellar dust which has formed in the expanding atmosphere of a cool star. Of
course there could be an involved evolutionary connection between the two
(§11.4).

The second class consists of the central objects in cometary reflection nebulae, some of which were mentioned (by CRL number) in Chapter 5.
Some work has suggested that a subset of these might represent an intermediate evolutionary stage between objects such as the IRC sources described
and planetary nebulae. Confirmation of this picture would be an important
step indeed, but considerably more optical, infrared, and radio data will
be needed first.

A survey of other types of source. Infrared excesses are often found
simply upon examination of the infrared spectrum of a previously known
object. The criterion that the object is seen optically selects somewhat
against highly obscured objects, so that the infrared excesses discovered
tend to be smaller than for the IRC and CRL sources. Objects from a wide
variety of classes are found to show infrared excesses, though not every
object within a certain class displays an excess.

The objects can be grouped according to whether the observed grains condense in the circumstellar gas or whether they are interstellar grains that have survived the evolution of the star to that stage. Prominent members of the first group are cool M and C stars. The oxygen-rich and carbon-rich separation described above is well established. It has been demonstrated that the existence of circumstellar dust and the strength of the excess is correlated with the coolness and the luminosity (extent) of the star. Total mass-loss rates are inferred from spectroscopic data to be 10^{-8} M_Θ y^{-1} or more. Many of the cool stars are variable, including Mira, semi-regular, and irregular long-period types. Such information will no doubt be important in understanding the condensation process.

Among normal stars of intermediate spectral type (F, G, and K) only the most luminous (class Ia) exhibit excesses. As an alternative to grain formation near these hotter stars it has been suggested that the dust may be 'fossil dust' remaining from an earlier M-supergiant phase of the evolution of these massive stars. In this context it is interesting that the Cepheid variable RS Puppis has several separate dust shells visible around it, a distinctive type of reflection nebula!

Added to the list of stars which generate their own dust are RV Tau stars (G-K bright giants with semi-regular variability and mass loss), R Cor Bor stars (F-G supergiants that are H-deficient, C-rich eruptive variables), the peculiar object η Carinae, planetary nebulae, and novae. Particularly interesting are novae in which decreases in optical flux correlated with increases in infrared flux have been observed, giving direct evidence for condensation of the dust. Similarly, new grains appear to be ejected near optical minima of R Cor Bor stars. The interpretation of the latter is not simple, however. Changes in the infrared flux without accompanying visual changes have occasionally been detected, while at other times the infrared flux remains constant when the optical flux decreases; these observations suggest an asymmetry or even patchiness in the dust distribution. This might be related to dust production in cool segments of coarse convective structure in the photospheres of these stars.

Naturally there has been considerable interest in young stellar objects associated with complexes of interstellar material. Many have significant infrared excesses, including T Tau stars and the early-type stars like them, Herbig Ae and Be stars associated with nebulosity, Herbig-Haro objects, and even non-emission-line 'Orion variables'. An interpretation in terms of dust remaining from earlier stages of pre-stellar collapse is appealing. Direct evidence for close association with dust comes from reflection nebulosity, most notably for some Herbig-Haro objects (Chapter 5) and those members which illuminate cometary nebulae.

P Cygni line profiles in the Ae and Be classes show that mass loss is a common phenomenon, as in T Tau stars; a typical outflow rate is 10^{-8}-10^{-7} $M_\odot$ y^{-1} at 200 km s^{-1}. Negative velocity shifts of some Herbig-Haro objects relative to surrounding dark-cloud material can also be interpreted as indicating mass loss, up to 10^{-7} $M_\odot$ y^{-1}. Survival of the remnant dust exposed to these stellar winds can be questioned introducing an alternative source of the hot dust, condensation in the outflowing material. It has also been argued that a strong correlation between Hα emission strength and infrared flux implies a free-free rather than a dust origin of the radiation. Against the free-free interpretation are observations of character-istic 10 and 20-μ 'silicate' emission in T Tau stars and cometary nebulae, as well as in the non-emission-line 'Orion variables'. Nevertheless it is possible, given the wide range in observed properties, that free-free emis-sion does contribute to some spectra, especially in the near-infrared (λ < 5 μ).

The related topic of thermal emission from dust in and around H II regions is taken up in §9.4.

7.2.2. Some properties of model circumstellar envelopes

Detailed models have been developed for circumstellar dust around cool stars where spherical symmetry may be used with most confidence. While we shall be describing these models in particular, much of the discussion is of a qualitative nature and can be carried over to the analysis of other circumstellar emission, including that from sources embedded in molecular clouds.

Optical depth. Although optically thin envelopes are most easily treated, it is important to ask whether they are in fact relevant. This can be judged by reference to observed spectra such as those for cool stars in Fig. 7.2. (For convenience they were plotted as λF_λ versus λ; a hori-zontal line on such a graph represents equal energy per decade.) Curves 1 and 10 give an impression of the underlying stellar spectrum, with little distortion from optical (near-infrared) absorption or excess infrared emis-sion. The other curves, in which the infrared emission is building up at the expense of the stellar optical radiation, represent cases of progres-sively increasing optical depth. When the envelope is optically thin the integrated infrared radiation L_{IR} is simply related to the stellar luminos-ity by

$$L_{IR} = \bar{\tau}_a L_*$$

(7.2)

where $\bar{\tau}_a$ is the effective optical depth of the envelope averaged over the stellar spectrum. (For a blackbody distribution $\bar{\tau}_a$ is the Planck mean.)

When $\bar{\tau}_a > 1$ proper treatment of the energy deposition in the shell requires a more sophisticated radiative transfer theory taking into account the albedo and phase function of the grains. Note that $\bar{\tau}_a > 1$ even for the upper curves in Fig. 7.2. Extreme cases are shown by curve 16, where the near-infrared radiation is severely diminished, and curve 9, where in addition strong self-absorption at 10 μ can be seen.

Temperature gradient. The grain temperature profile in the envelope is determined by a combination of the energy deposition and the escape of thermal radiation from the envelope. The problem is somewhat simplified when $\tau_a(IR) < 1$, even when $\bar{\tau}_a > 1$, since then reabsorption of thermal radiation can be ignored. Note also that when $\tau_a(IR) < 1$ the infrared spectrum can be computed easily once the temperature profile is known (eqn (7.4)). The radial dependence of the dust temperature in an optically thin envelope $(\tau_a(IR) < \bar{\tau}_a < 1)$ is easily calculated if the wavelength dependence of the absorption cross-section is known. Suppose again for the purpose of illustration that $Q_a \propto \lambda^{-n}$ throughout the spectral range of interest. Then by energy balance, as discussed in §6.2,

$$T_d(R) = T_*(R_*/2R)^{2/(n+4)} .$$

(7.3)

To provide some perspective Fig. 7.3 shows values of $Q_a(\lambda)$ actually used for the model computations discussed below. The three curves are for graphite ($a=0.05$ μ), and for 'dirty' and 'clean' silicates ($a=0.1$ μ). (The meaning of the adjectives dirty and clean becomes clear upon examination of

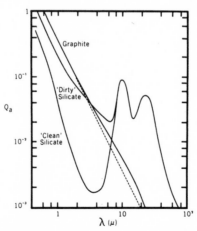

FIG. 7.3. Behaviour of Q_a for small spheres of graphite ($a=0.05$ μ) and silicate ($a=0.1$ μ). Broken curve is a λ^{-2} approximation for graphite; the dependence for silicates is much more complex, showing features near 10 and 20 μ.

the curves.) For graphite n≈2, as indicated, but for silicates, which have
dominant 10 and 20-μ features, the energy balance has to be treated numerically. The temperature profiles for the optically thin limits are shown in
Fig. 7.4 (broken curves). For graphite eqn (7.3) with n=2 is an excellent
approximation. The 'dirty' silicate profile is also a power law but is
steeper with n≈0. 'Clean' silicates would be somewhat cooler still because
their optical and near-infrared absorption is relatively low.

Eqn (7.3) can be used to place a lower limit on the inner radius of
the grain distribution, since T_d should not exceed the condensation temperature T_c if grains are forming around the star (or the melting temperature
if pre-existing grains are being heated). For graphite $T_c \simeq$ 1700 K, while
for silicates $T_c \simeq$ 1200 K. In the models to be described $T_* =$ 2400 K, so
that for graphite using n=2, $R/R_* >$ 1.4. For silicates it would be somewhat larger. (Values of 1.6 and 4 respectively were used in the models.)

Examination of Fig. 7.2 shows that even the simplification τ_a(IR) < 1
cannot always be used. When the shell becomes optically thick, the temperature profile changes for two reasons. Near the inner edge T_d rises if the
albedo is non-zero, since multiple scattering increases the probability of
absorption there. At the outer edge the temperature is lower, for although
the bolometric flux is the same the radiation has been thermalized to lon-

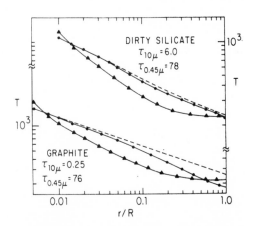

FIG. 7.4. Temperature distribution in model 'dirty' silicate
and graphite envelopes with outer radius $R_o = 2.4 \times 10^{16}$ cm (stellar temperature 2400 K). The broken curves are the limits at
zero optical depth. The other curves correspond to the optical
depths shown, with uniform (dots) or R^{-2} (triangles) density
distributions. (From T.W. Jones and K.M. Merrill, Model dust
envelopes around late-type stars, Astrophys. J. 209, 509
(1976). University of Chicago Press. ©1976 by the American
Astronomical Society. All rights reserved.)

ger wavelengths where it cannot be absorbed as efficiently. These effects
are apparent in the results displayed in Fig. 7.4.

Spectrum. Consider now what determines the shape of the spectrum of
thermal radiation (to which is added the obscured stellar radiation). For
simplicity let $\tau_a(IR) < 1$, so that the contributions of each layer of the
envelope can be computed independently and then summed. Such a contribution to the luminosity is

$$dL(\lambda,R) = n_d(R) \ C_a(\lambda) \ 4\pi B(T_d,\lambda) \ 4\pi R^2 \ dR \ . \qquad (7.4)$$

All other parameters being constant, the contribution is proportional to
mass, as would be expected.

In Fig. 7.5 curves in the form λF_λ for $C_a \propto \lambda^{-n}$ are shown for a shell
with $T_d = 300$ K. The vertical scales are adjusted so that each curve corresponds to the same L_{IR}; thus the shells have equal $\bar{\tau}_a$ as well as temperature. Notice how the energy distribution shifts to shorter wavelengths as
n increases (cf. eqn (7.1)). The long-wavelength tail falls as λ^{-n-3} in a
λF_λ plot. When $C_a(\lambda)$ has detailed structure, like that for silicates, corresponding structure, as distorted by $B(T_d,\lambda)$, will occur in the emitted
spectrum.

Consider a second larger shell which has the same optical depth and

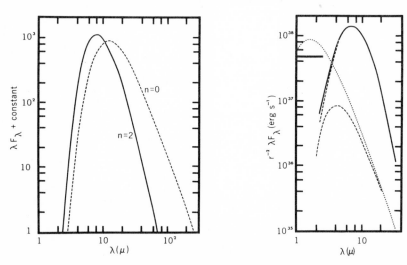

FIG. 7.5. Spectra for optically thin shells with $T_d = 300$ K,
opacity dependence λ^{-n}, and identical bolometric luminosities.

FIG. 7.7. Model emergent spectra of a shell of graphite
grains and the underlying star, in the format of Fig. 7.6.

L_{IR} as the first. Since the temperature is lower the spectrum is shifted to longer wavelengths. If C_a maintains the same functional form in the new wavelength regime, then the shift in Fig. 7.5 is simply horizontal, in proportion to T_d^{-1}. The shape of the spectrum in a real envelope, in particular the amount of broadening, clearly depends on the range of radii, the temperature gradient (eqn (7.3)), and $n_d(R)$, which specifies among other things where the optical depth is concentrated in the envelope.

Some of these effects can be illustrated with power-law representations of the density, $n_d(R) \propto R^{-m}$. When m is large the major contributions to optical depth and luminosity occur near the inner edge. The spectrum is correspondingly narrow, like that for a single temperature. As m decreases the spectrum broadens. For example, when m=1 equal logarithmic intervals in R make the same contribution to $\bar{\tau}_a$ and hence to L_{IR} when $\bar{\tau}_a < 1$, but the emission is shifted to increasingly long wavelengths as described. At sufficiently long wavelengths $B_\lambda \propto T_d$; using eqns (7.3) and (7.4) it can be shown that the long-wavelength emission is dominated by the outer cooler grains (the spectrum is broadened) when $m < 3-2/(4+n)$, as will usually be the case. The hottest grains always dominate the short-wavelength emission. On the long-wavelength side of the broadened peak λF_λ eventually falls off as λ^{-n-3} for any m. It is perhaps worth remarking that for a given L_{IR} or optical depth most efficient use of mass is made by central concentration, since the mass is a stronger function of R than is optical depth. Grains at smaller radii individually receive a greater stellar flux, and so are hotter and radiate more.

We have seen that changes in grain size and composition affect the grain temperature. Therefore additional broadening of the spectrum can arise if various grain species coexist within the envelope.

<u>Emission or absorption features</u>? Some materials have absorption bands in the infrared; silicates and silicon carbide have been mentioned. Suppose such a marked increase in C_a occurs in a band at λ_B. Let $T_B \simeq 0.3/\lambda_B$ be the Wien-law temperature at the band. If $T_d < T_B$ the grains are unable to emit significant amounts of radiation in the band and so will only absorb against the underlying continuum. This continuum may be the stellar continuum, especially if $\tau_a(\lambda_B) < 1$, or may be that emitted by hotter dust closer to the star. If it is composed of the same material, any hotter dust for which $T_d > T_B$ can produce an emission feature in the spectrum against which the cooler dust is absorbing. Clearly the resulting feature can be either in emission or in absorption, depending on the relative amounts of hot and cool dust. If the observed spectrum is to be used to derive quantitative information on the amount of material present and on the intrinsic behaviour of $C_a(\lambda)$, which helps determine the composition, a

proper transfer solution would be needed.

In the infrared sources being considered the amount of near-infrared continuum emission indicates the presence of dust hot enough to produce a 10-μ feature in emission ($T_d \simeq 300$ K). Some cool stars, however, have sufficient cooler dust in the outer layers of their circumstellar envelopes to produce a net absorption, as is apparent in Fig. 7.2. Notice also how the 20-μ feature, for which the optical depth is smaller and the critical temperature is a factor of 2 lower, tends to remain in emission. Cold dust in the interstellar medium can also cause absorption features. The striking 3.1-μ ice and 10-μ silicate absorption in the BN object in Orion has already been mentioned as a possible example. If all the dust is confined to a molecular cloud which is progressively hotter towards the centre, then the cold dust should be treated in a unified radiative transfer model rather than as a cold foreground screen.

Models. To supplement this qualitative description, the results of actual radiative transfer calculations over a considerable range of model parameters are shown in Fig. 7.6. Such a grid of models is useful in narrowing down the choice of parameters to fit the observations of a particular star; even so it is difficult to explore the whole range of possibilities. This particular grid is based on the 'dirty' silicate particles for which $Q_a(\lambda)$ was shown previously (Fig. 7.3). The underlying star has a blackbody spectrum with temperature 2400 K. To satisfy the condensation temperature condition an inner radius $R_i = 4R_*$ was chosen. Shells of two thicknesses, given by R_o/R_i equal to 20 and 100, were studied; results for the more extended envelope appear in the lower half of the panels. Within the envelope power-law density distributions with m equal to 0, 1, and 2 were used (as labelled). Finally the parameter τ (as scaled from τ(10 μ) which is shown) was varied from model to model. The horizontal bar on each curve indicates the vertical displacement needed for consistent scales of λF_λ. Note that to this point R_* has not been specified. If $L_* = 3.25 \times 10^4 r^2 L_\odot$, then $R_* = 7.3 \times 10^{13} r$ cm (1050 r $R_\odot$) and the horizontal bar is $5 \times 10^{37} r^2$ erg s^{-1}. Temperature distributions for a few of the models were shown in Fig. 7.4. Recall that if several types of grain were included in the model, then the temperature distribution for each would be computed separately.

Fig. 7.6 will not be discussed in detail. However, if it is inspected systematically, changing one parameter at a time, many of the qualitative effects that were described above will be seen. Similarly, comparison of the grid of models with the range of observed spectra (Fig. 7.2) shows that many of the features and trends can be explained.

A second grid of models for graphite particles has also been com-

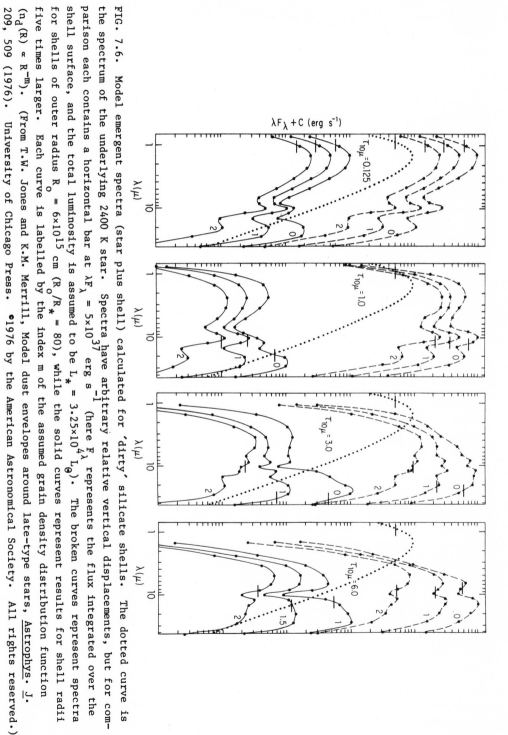

FIG. 7.6. Model emergent spectra (star plus shell) calculated for 'dirty' silicate shells. The dotted curve is the spectrum of the underlying 2400 K star. Spectra have arbitrary relative vertical displacements, but for comparison each contains a horizontal bar at $\lambda F_\lambda = 5 \times 10^{37}$ erg s^{-1} (here F_λ represents the flux integrated over the shell surface, and the total luminosity is assumed to be $L_* = 3.25 \times 10^4 L_\odot$). The broken curves represent spectra for shells of outer radius $R_o = 6 \times 10^{15}$ cm $(R_o/R_* = 80)$, while the solid curves represent results for shell radii five times larger. Each curve is labelled by the index m of the assumed grain density distribution function $(n_d(R) \propto R^{-m})$. (From T.W. Jones and K.M. Merrill, Model dust envelopes around late-type stars, <u>Astrophys</u>. <u>J</u>. 209, 509 (1976). University of Chicago Press. ©1976 by the American Astronomical Society. All rights reserved.)

puted. Fig. 7.7 displays the relationships between the spectra of the
thermal emission, the unreddened central star, the reddened star, and the
total emission for one member of the grid ($R_i/R_* = 1.6$, $R_o/R_i = 50$, m=0,
$\tau(10~\mu) = 0.25$). (Actually the flux from the envelope is about 20 per cent
too high near the peak because for the purposes of our computation we
approximated the temperature distribution with a simple power law which is
not quite self-consistent. This shortcoming is of little consequence for
the discussions below.)

   Size. Data on the spectral dependence of flux and the bolometric
luminosity are the most commonly available for comparison with theoretical
models. However, the importance of measurements of angular size as a function of wavelength should not be overlooked as further restrictions on the
model parameters are sought. Measurements of angular size have been made
for a few sources during lunar occultations. Several others have been
observed using two-element Michelson interferometers, a technique that in
principle can be extended to many more sources. Multi-aperture photometry
is an alternative for the most extended sources. Simplistic interpretations in terms of uniformly bright disks can be offered. However, model
dust envelopes can also be used to predict the wavelength dependence of the
angular size. Some general statements can be made. First, if the dust is
confined to a shell that is both optically and geometrically thin, then the
size should not depend on wavelength. On the other hand, both optical-
depth effects and the temperature gradient present in an extended envelope
can be expected to produce a change in size with wavelength. Second, a
source should appear no larger than the radius at which $T_d(R)$ equals the
Wien temperature for the observed wavelength, since for smaller T_d (larger
R) dust does not emit efficiently. Usually the size is considerably smaller than this. Third, the angular size at the wavelength of a self-absorbed spectral feature should be larger than that in the adjacent continuum. This would help distinguish such sources from those obscured by
intervening but unrelated interstellar dust.
   To be specific the brightness distribution of the model in Fig. 7.7
has been computed. At the wavelengths of interest ($\lambda > 2~\mu$) the particle
albedo is low (<0.01) and so the emission may simply be integrated along
rays of constant impact parameter, taking into account the optical depth
from each contributing point to the surface. Through the envelope the
temperature distribution was approximated by $T_d = 1625~(R_i/R)^{0.42}$ K, following that for the more extended model shown in Fig. 7.4. (The power 0.42
may be a slight underestimate but will suffice.) Recall that the effects
of reabsorbed thermal radiation were included in determining the temperature distribution that we have approximated.

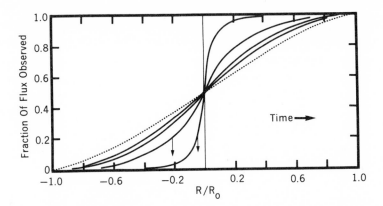

FIG. 7.8. Reappearance curves for the dust envelope of Fig. 7.7, at 2.2 (steepest), 4.9, 9.9, and 27 μ (the broken curve is for a uniformly bright disk). The arrows point to the radius at which T_d is the appropriate Wien temperature (for the two longest wavelengths this radius is not inside the envelope).

To model what is observed during a lunar occultation of an extended infrared source the strip brightness distribution can be integrated over the portion visible, which varies with time. Reappearance curves (for the dust component only) at several wavelengths are plotted in Fig. 7.8, each normalized to unit total brightness. For reference the arrows point to the radius at which T_d is the appropriate Wien temperature. The brightness distribution for a uniform disk is also shown; even with a change in scale it is a poor approximation to the real curves as it lacks the extended tail. In this model the obscured star begins to dominate at $\lambda < 2$ μ, and so the angular size of the thermal dust source becomes irrelevant to observations. However, because the albedo increases towards shorter wavelengths the small stellar disk may not actually be seen if there is enough scattering to produce a relatively extended reflection nebulosity or halo.

The fringe visibility can be computed from the Fourier transform of the strip brightness distribution for comparison with measurements by interferometry. Some visibility curves for the above model are displayed in Fig. 7.9. The scale for the interferometer spacing S corresponds to an outer angular radius of 1 arc sec (e.g. r=1 and a distance of 400 pc). As a point of reference we observe that the 27-μ visibility of a uniformly illuminated disk of the same size would have its first null at S=340 cm and would not be too different from the 27-μ curve shown. The behaviour of the visibility as the wavelength decreases may be interpreted as the effect of shrinking source size or increasing limb darkening. (The effects of changing resolution could be removed by using S/λ as abscissa.) Note that the contribution of the central source, which would be important by 2.3 μ, has

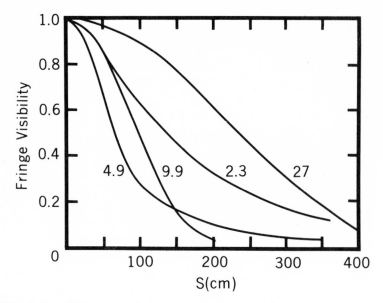

FIG. 7.9. Fringe visibilities for the model of Figs. 7.7 and
7.8.

been omitted here.

    Polarization. Many late-type stars with circumstellar envelopes also
show intrinsic linear polarization at optical and near-infrared wavelengths. Usually this has been interpreted as the result of scattering by
dust in the envelope, although more recent evidence suggests that patchiness in the stellar photosphere might contribute in some cases. If the
scattering interpretation is correct, either the illumination or geometrical arrangement of the dust must be asymmetrical. Wavelength dependence of
the position angle, which is commonly seen, can be attributed to the additional effects of optical depth. Circular polarization found in some
sources may originate in multiple scattering, but no more than a qualitative treatment has been given. Time variability is also prominent.
    Modelling the polarization is in principle like that for reflection
nebulae (§5.2). However, for circumstellar envelopes there is an important
difference, the lack of any observational data on the spatial dependence of
polarization. Therefore the task is more difficult and considerable
ambiguity remains.

FURTHER READING

§7.1
FAZIO, G.G., and STECKER, F.W. (1976). Prediction of the diffuse far-in-

frared flux from the galactic plane. Astrophys. J. Lett. 207, L49.

SERRA, G., PUGET, J.L., RYTER, C.E., and WIJNBERGEN, J.J. (1978). The
    far-infrared emission of interstellar matter between longitudes ℓ=36°
    and ℓ=55°. Astrophys. J. Lett. 222, L21.

§7.2.1

CODE, A.D. (1973). Radiative transfer in circumstellar dust clouds. In
    Interstellar dust and related topics (eds. J.M. Greenberg and H.C.
    van de Hulst). Reidel, Dordrecht.

MERRILL, K.M., and STEIN, W.A. (1976). 2–14 μm stellar spectrophotometry.
    I Stars of the conventional main sequence. II Stars from the 2 μm
    Infrared Sky Survey. III AFCRL Sky Survey objects. Publ. astron.
    Soc. Pac. 88, 285, 294, 874.

NEY, E.P., and HATFIELD, B.F. (1978). The isothermal dust condensation of
    Nova Vulpeculae 1976. Astrophys. J. Lett. 219, L111.

LEE, T.A. (1973). Visual and infrared photometry of RY Sagittarii near the
    phase of deep minimum. Publ. astron. Soc. Pac. 85, 637.

See §5.1. (Strom, Strom, and Grasdalen 1975).

§7.2.2

SCOVILLE, N.Z., and KWAN, J. (1976). Infrared sources in molecular clouds.
    Astrophys. J. 206, 718.

JONES, T.W., and MERRILL, K.M. (1976). Model dust envelopes around late-
    type stars. Astrophys. J. 209, 509.

McCARTHY, D.W., LOW, F.J., and HOWELL, R. (1977). Angular diameter mea-
    surements of σ Orionis, VY Canis Majoris, and IRC +10216 at 10.2 and
    11.1 micrometers. Astrophys. J. Lett. 241, L85.

SHAWL, S.J. (1974). Polarimetry of late-type stars. In Planets, stars and
    nebulae studied with photopolarimetry (ed. T. Gehrels). University
    of Arizona Press, Tucson.

KRUSZEWSKI, A., and COYNE, G.V. (1976). Wavelength dependence of polariza-
    tion, XXXI. Cool stars. Astron. J. 81, 641.

8

INTERACTIONS IN AN INTERSTELLAR ENVIRONMENT

Interactions involving grain surfaces and the dynamics of interstellar grains are of fundamental importance to a wide variety of phenomena. Unfortunately many are sufficiently sensitive to changes in the grain material, surface conditions, and the interstellar environment that quantitative predictions cannot be made with certainty. We shall introduce some basic processes that have to be considered, and shall estimate their potential importance in specific situations. In other applications the relative importance of contributing factors should be evaluated anew. Further laboratory work will be necessary for a complete understanding.

8.1. PHENOMENA ON GRAIN SURFACES

Binding of atoms and molecules to the grain surface. Atoms and molecules are only weakly bound to grain surfaces through van der Waals interaction. Some examples of the low binding energies D that characterize physical adsorption are given in Table 8.1. The existence of pure hydrogen grains in the interstellar medium is precluded by the weak binding of H_2. Other pure molecular grains would be more tightly bound, with D/k as high as 6200 K for pure H_2O ice. However, such regular crystals are unlikely to develop by accretion because at the same time many other species, such as C and N, would be striking the surface. It is not necessary to discuss valence bonding ($D/k > 3.5 \times 10^4$ K $\simeq$ 3 eV) since a highly reactive surface would soon be covered with sufficient chemisorbed material to render it inactive; chemisorption is so strong that at typical grain temperatures evaporation cannot reverse the process.

Grain surfaces are undoubtedly not regular lattice structures. Impurity atoms, growth ridges, dislocations produced by incident X-rays, ultraviolet photons, and low-energy cosmic rays, and indeed any concave feature in the surface can contribute to the phenomenon called 'enhanced

TABLE 8.1
EXPERIMENTAL AND ESTIMATED DESORPTION ENERGIES (K)

Species	Surface					
	H_2O	CO_2	CO	H_2	'Dirty' graphite	'Regular inert'
H	690	640	280	(100)	785	
H_2	860	800	350	100	980	
CO	(2500)	1950	1000			
H_2O	6200					(1000 – 2000)
C,N,O						(800 – 1200)

binding', including 'semi-chemisorption'. As will be discussed below, this concept may be particularly significant for H_2 formation.

Sticking probability. The impact energy, equivalent to the kinetic energy and hence about 100 K in a typical H I region, is much less than D/k but exceeds the energy corresponding to thermal evaporation. In these conditions a hydrogen atom, being much lighter than the atoms or molecules which form the grain surface, transfers little of its energy in one collision. Therefore it bounces along the surface several times before sticking; sometimes it will escape before it is thermally accommodated. Both theory and experiment indicate that for hydrogen impacts on a regular surface the sticking efficiency S is about 0.3. A variety of surface imperfections are expected to increase the efficiency. Data for H_2 indicate $S \simeq$ 0.7, increasing with the number of H_2 monolayers. Heavier atoms stick with unit efficiency because their collisions are more inelastic and the binding force is greater. The sticking of low-energy electrons has also been studied in the laboratory. Here the energetics are somewhat dependent on the electric potential of the surface, but it appears that electrons should also stick with high efficiency.

These high sticking probabilities point to the possible importance of three related phenomena: the freezing out of a large fraction of interstellar gas on grains, the consequent growth of grain mantles, and molecule formation on grain surfaces as a source of molecules in the interstellar medium. There are of course other conditions to be met, which we shall discuss.

Time scales for gas-grain interactions. One prerequisite is that the cool H I cloud environment in which the grain is embedded remains reasonably stable for a time long enough for the cumulative effects on gas and grains to become significant. If the H I gas becomes hotter, say in the shock of a violent cloud-cloud collision or in the diffuse dissipated cloud, sputtering away of the fragile mantle could effectively undo these processes. The time between catastrophic collisions can be estimated from typical cloud sizes, velocities, and space densities to be

$$t_{cc} \simeq 3 \times 10^{15} \text{ s .} \tag{8.1}$$

Less violent collisions, occuring perhaps ten times more frequently, could seriously reduce the molecular abundances.

Gas-grain interactions proceed more rapidly in dense clouds. However, these same clouds are often also regions of star formation; when a star finally forms the grains can be disrupted. A lower limit to the time

required for star formation is the free-fall time for the most dense
regions of the cloud complex,

$$t_{ff} \simeq (3\pi/32 \; G \; m_H n_H)^{\frac{1}{2}} \simeq 1.6 \times 10^{15} \; n_H^{-\frac{1}{2}} \; s \qquad\qquad (8.2)$$

which is comparable with t_{cc}.

There are two distinct interaction time scales. If we are interested
in the effects on a single grain, then we wish to know the rate at which
gas atoms strike its surface. For a gas species at $T_g \simeq 100$ K the characteristic time is

$$t_s = (n_a \; \sigma_d \; <v_a>)^{-1} \simeq 5.7 \times 10^3 \; n_H^{-1} \; (n_H/n_a)(m_H/m_a)^{\frac{1}{2}} \; s \; ; \qquad (8.3)$$

the latter expression corresponds to a grain radius of 0.2 μ. A closely
related time is that for the grain to be struck by the equivalent of a monolayer. If there are N lattice sites on the surface, with dimension ℓ,
then

$$t_{mon} = N t_s = 4(\ell^2 \; n_a \; <v_a>)^{-1} \simeq (3\ell/a) \; t_m \qquad\qquad (8.4)$$

where 3ℓ/a is the monolayer to total volume ratio and t_m is the time in
which the grain is struck by its own mass (eqn (8.18)). If we take $a \approx 0.2$ μ
and $\ell \simeq 6$ Å, then $N \simeq 10^6$ and $3\ell/a \simeq 10^{-2}$. Therefore $t_s \ll t_{mon} \ll t_m$. Even
t_m (Table 8.2) is short compared with t_{cc} and t_{ff}.

To study whether the interstellar gas becomes depleted through condensation onto grains or if there is a significant processing of the gas
into molecules via surface recombination, we need to evaluate a different
time scale, that required for a given gas atom to stick to any grain surface. Clearly

$$t_{dep} = (n_d \; \sigma_d \; <v_a>)^{-1} = (n_H/n_d) t_s (H) \simeq 3 \times 10^{16} \; n_H^{-1} \; (a/0.2 \; \mu) \; s \; ; \quad (8.5)$$

for the numerical result we have assumed that a large fraction of the heavy
atoms is in grains. Though much larger than t_s and the other related time
scales, t_{dep} is still smaller than t_{cc} and t_{ff} if $n_H > 10$ cm^{-3} and 100
cm^{-3}, respectively, as will usually be the case.

Evaporation time and critical temperatures. Having seen that the
gas-grain interactions are potentially important, we must investigate
whether evaporation limits the cumulative effects. Since thermal evaporation results from transfer of vibrational energy to surface atoms, the
characteristic escape time is

$$t_{ev} \approx \nu_o^{-1} \exp(D/kT_d) \qquad\qquad (8.6)$$

where ν_o is the vibrational frequency of the lattice, typically 10^{12} Hz.
Since $D \gg kT_d$ even for the weakest bonding we have considered, t_{ev} is very
sensitive to changes in D and T_d. If the surface is chemically active it
will acquire a relatively inert monolayer in a time t_{mon}. Because D is so
high the evaporation of this monolayer is quite negligible. The type of
attachment of further atoms on this monolayer depends on whether a regular
crystal structure is preserved in the monolayer. Even with adequate surface migration this seems unlikely because of the variety of atoms arriving
at the surface. Thus adhesion to the monolayer is probably weak, as outlined in Table 8.1. To judge the significance of various processes we must
compare t_{ev} with t_s, t_{mon}, and t_{dep}; each result can be summarized as a
grain temperature above which evaporation is dominant.

For example if an atom remains on a grain long enough for a second
atom to stick to the grain, then there is a possibility of molecule formation; the criterion is $t_{ev} \geqslant t_s$, with the equality defining $T_s = T_d$. As
consequences of the large exponential argument in eqn (8.6), T_s is not very
sensitive to our choice of parameters and T_d need be only slightly smaller
than T_s for evaporation to become negligible. The critical temperature T_s
is plotted as function of D in Fig. 8.1. It was evaluated for $t_s = 570$ s,
as obtained from eqn (8.3) for hydrogen with $n_H = 10$ cm^{-3}. Similar considerations give the curves for T_{mon} and T_{dep}. Here we used $t_{mon} = 5.7 \times 10^8$ s
from eqn (8.4) with $N=10^6$, and $t_{dep} = 3 \times 10^{15}$ s from eqn (8.5).

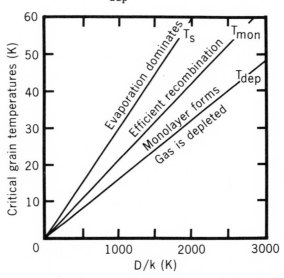

FIG. 8.1. Critical grain temperatures as a function of the
binding energy at the grain surface.

We note parenthetically that the uncertainties in evaluating the total evaporation rate from eqn (8.6) can be effectively circumvented for materials for which appropriate low-temperature equilibrium vapour-pressure data are available.

Surface mobility. The movement of an adsorbed atom or radical is restricted by the potential barriers of the regular lattice structure and impurity sites. The time scale for a thermal hop to an adjacent site has the same form as t_{ev}, with D replaced by the barrier height E_b; the time to sample N sites is N^2 times longer. On an inert surface with physical adsorption E_b is typically $\frac{1}{2}D$, except for H for which E_b can be a smaller fraction of D. Thus thermal diffusion over the surface can be quite rapid relative to the rate of evaporation. Of equal or greater importance is quantum-mechanical tunnelling through the barrier. This is particularly efficient for hydrogen, which can migrate over an icy grain surface in a small fraction of a second. Surface mobility is generally so high that it is not a concern in discussing the rate of other physical processes which involve the motion of particles across the surface. For example, every atom sticking to the surface has an opportunity to form a molecule when $T_d < T_s$; even if a particular atom becomes relatively tightly bound and immobile it is soon 'visited' by a second.

Molecule formation. When $T_d < T_s$ there is efficient recombination of atoms sticking to the surface. This is pointed out in Fig. 8.1. Whether the molecule formed remains on the surface is a separate problem. For simple unsaturated radicals like CH, NH, and OH it is possible that stronger binding like chemisorption holds them on the surface. However, subsequent recombination with hydrogen will eventually lead to CH_4, NH_3, and H_2O, which are much more likely to be ejected during formation. Unfortunately the fraction of heavy atoms that ultimately remains as part of a hydrogen-saturated surface is quite uncertain, estimates ranging from 0.01 to 0.8. If the low estimate is correct then interstellar grain surfaces are good sources of interstellar molecules. Of course to determine the equilibrium gas-phase abundances of molecules, molecular destruction (chiefly by ultraviolet dissociation) would have to be considered.

H_2 formation warrants special attention. Ignoring CO surfaces for the moment, we see from Table 8.1 and Fig. 8.1 that $D/k \simeq 700$ K, so that $T_s \simeq 20$ K. This upper limit for T_d is not very restrictive. What happens if $T_d < T_{mon}$? Experimental data show that as a surface becomes covered with successive monolayers of H_2 the adsorption energy of H_2 decreases. As few as 20 layers are sufficient to reduce the value to 100 K, appropriate for a pure H_2 surface. The build-up of H_2 monolayers is thus self-regulating; it should cease when D has fallen to the value that makes $T_{mon} \simeq T_d$. (Conversely, a continued growth of pure hydrogen mantles beyond a few monolayers would require $T_d \simeq 2$ K. The cosmic microwave background radiation alone will keep interstellar grains warmer than that, and so pure hydrogen grains do not exist. That explains why the mass fraction of interstellar material locked in solid particles does not exceed about 1 per cent.)

The value of D for H on the coated surface will be lower as well, thus lowering T_s. However, as long as $D(H) > 0.7 D(H_2)$, as seems to be the case for the examples in Table 8.1, then $T_s > T_d$ when $T_d \simeq T_{mon}$ (Fig. 8.1). Therefore H_2 molecule formation should not be inhibited.

If in a dense cloud the CO abundance were to exceed about $10^{-5} n_H$, then grains could become coated with CO rather than H_2, thus lowering T_s for H to about 9 K (Fig. 8.1). Methane and ammonia might produce a similar effect. In light of the widespread presence of H_2 any such value of T_s is uncomfortably low. This brings in the concept of 'sites of enhanced binding'; an increase in D/k to only 500 K would raise T_s to 15 K. In addition to the above-mentioned possibilities for enhanced binding, it has been suggested that the formation of CO_2 from incident O atoms might provide appropriate sites. The energy of formation of H_2 is about 5×10^4 K (4.48 eV), much in excess of the binding energy even at an enhanced site. Therefore transformation of only a small fraction into translational energy would be sufficient for ejection. An estimated ejection probability in excess of 0.9 is supported by laboratory studies. (Thus H_2 monolayers must be built up from gas-phase molecules that strike the grain.) The fraction of the recombination energy that is transferred to the grain, and hence could contribute substantially to heating the grain, is quite uncertain, but laboratory measurements suggest it is small.

A wide variety of grains in different environments, including unshielded regions, may thus contribute to H_2 formation. The equilibrium fraction of H in molecular form is enhanced in regions shielded from ultraviolet radiation for several reasons: T_d is lower and so most grains will be colder than T_s; H_2 is self-shielding against ultraviolet dissociating radiation, making the destruction rate lower; the shielded regions in question are likely to be more dense (rather than larger) than average clouds, so that the production rate is higher. Comparison of theoretical equilibrium abundances with astronomical observations suggests a formation efficiency near unity.

Growth of grain mantles and depletion of the interstellar gas. When $T_d < T_{dep}$ thermal evaporation is not sufficient to prevent the interstellar gas from condensing on grain surfaces. Even for weak binding (D/k $\simeq$ 500 K, $T_{dep} \simeq 8$ K) suitable conditions can be found in shielded regions in the

interior of dense dust complexes. This is not a sufficient criterion, however, since atoms that stick soon form molecules which may escape during
the recombination process. If a fraction δ of the molecules is retained in
successively built up monolayers, then deposition of the gas proceeds on a
time scale t_{dep}/δ. This is less than t_{cc} and t_{ff} when $n_H > 10/\delta$ cm^{-3}.
Since it is unlikely that δ is as small as 10^{-2}, this condition can be met
in dark clouds (that are already required for low T_d). Grains cannot grow
indefinitely because there is a limited supply of gas; for example if
grains normally take up 25 per cent of the maximum available, then an
increase in grain size by mantle growth in a dark cloud is at most by a
factor of 1.6. What condensable material remains in the gas phase is
largely in molecular form.

Direct evidence bearing on this question is found in the ρ Oph dark
cloud complex. There it is found that stars with larger E_{B-V} in the more
obscured regions have larger values of both λ_{max} and E_{V-R}/E_{B-V} (R) as well.
This suggests that grain-mantle growth has indeed occurred. The range in
λ_{max} is 0.55-0.81 μ, consistent with size changes by up to a factor of 1.5.
Coagulation of grains might also explain the observations, however
(§11.3.2).

Non-thermal evaporation - photodesorption. When $T_{mon} < T_d < T_s$ atoms
sticking to the surface form molecules, but thermal evaporation is sufficient to prevent the build-up of successive monolayers. However, at lower
temperatures mantle growth will occur if even a small portion δ of molecules remains on the surface after formation. Any further mechanisms which
can return molecules to the gas phase and hinder the growth of grains are
therefore important. Many have been evaluated, including the effects of
soft X-rays, low-energy cosmic rays, and infrared radiation; quantitatively
these prove to be uninteresting.

On the other hand, it is found that ultraviolet photons efficiently
remove physisorbed particles. Underlying the ejection are several different possible processes. A particle can be electronically excited to an
unbound state in the particle-surface potential, from which it can escape
if it is not de-excited in a time $t \approx \nu_o^{-1}$, or a molecule absorbing a photon
might actually be dissociated, and the fragments may then escape with the
assistance of the excess kinetic energy, or may electronically dissociate
if they are in excited electronic states, or may escape after recombination
into a vibrational-rotational excited state if the energy of this excitation is converted into translational energy on collision with the surface.
Because of the variety of possible branches leading to ejection it is
believed that for physical adsorption (the type of binding in which we are
most interested) the probability of photodesorption is high. A limited

amount of experimental evidence is available to support this. The yield
per photon incident on the grain is nevertheless not unity, since only the
surface monolayer is involved.

The maximum desorption is expected at wavelengths where the gas-phase
species is strongly absorbing, though the spectrum of the adsorbed species
is not identical. The radiation field of interest is therefore from about
6 to 13.6 eV. As a guide to the time scale in which a surface particle is
removed we can use estimates of the lifetimes of interstellar molecules,
typically 10^9 s in unobscured H I regions. When $n_H \simeq 10$ cm^{-3} and C, N, and
O with H make up the monolayer, $t_{mon} \simeq 2 \times 10^{12}$ s. Thus t_γ is short compared with t_{mon} and only a fraction $\delta t_\gamma / t_{mon} \simeq \delta\, 10^{-3}$ of the surface is
covered with molecules. Since there is considerable uncertainty in the
photodesorption process, these numbers should be used with caution. It
does appear though that optical erosion is an important process, since a
monolayer can be removed so quickly. Photodesorption in an H II region
should have an even more pronounced effect.

The destructive process would be limited if there were more cohesion
of the surface than is provided by physical adsorption. This might actually be promoted by the effects of absorption of ultraviolet photons
(§11.2). Furthermore, in more dense H I regions obscuration severely limits the ultraviolet photon flux. The amount of radiation that does leak in
depends on the albedo and phase function of the grains. If there were a
magnitude of visual extinction then t_γ would rise to at least 10^{10} s, while
t_{mon} would fall to 2×10^{10} s for $n_H \simeq 10^3$ cm^{-3}. Thus the regions most
favourable for rapid grain growth and depletion of the gas are also prob-
ably unaffected by photodesorption. At the same time this creates a potential problem for processing gas atoms into molecules using surface reactions, because then the only apparent way left to remove the molecules is
ejection at the time of recombination.

Sputtering. An atom striking the surface of a grain with sufficient
energy can knock atoms or molecules from the surface layers. Consider the
following formula for the energy threshold for sputtering:

$$E_{th} = \{(m_g + m_t)^2 / (1 + \alpha)^2\, m_g m_t\}\, G\, D . \qquad (8.7)$$

The first factor is the energy-transfer factor including possible inelastic
sputtering ($\alpha < 1$); since the target particle is usually the more massive
($m_t > m_g$) this factor exceeds unity. G (> 1) expresses the strong functional dependence on the angles of incidence and ejection. Thus E_{th} is
considerably greater than D. In a thermal gas sputtering proceeds efficiently when $T_g > E_{th}/k$. Sputtering of refractory grains (high D) clearly

requires more extreme conditions than sputtering of icy particles or weakly adsorbed molecular monolayers. Note also that at a given temperature there can be selective sputtering depending on D. Sputtering is negligible in the characteristically low-temperature environment of an H I region.

The rate of sputtering is $n_g v_g Y_g \pi a^2$ where Y_g is the yield per impact of gas species g. The yield is a function of geometry and of energy and so the integration of $v_g Y_g$, taking account of a Maxwellian distribution of gas velocities, systematic drift of the particle with respect to the gas, and the change in cross-section for charged-grain ion encounters, is quite complicated. This integration has been treated theoretically but still contains a constant of proportionality that has to be determined by experiments. Some predicted normalized yields, $Y_g' = \langle v_g Y_g \rangle / \langle v_g \rangle$, are sketched in Fig. 8.2 as a function of gas temperature for a selection of drift velocities. Two dominant interactions in a hot neutral gas are shown, He on H_2O (D/k assumed, 3250 K) and H on CH_4 (D/k assumed, 700 K).

Sputtering reduces the grain radius on a time scale

$$t_{sp} = a(da/dt)^{-1} = a\{(m_t/4s_d)(n_g/n_H) \langle v_g \rangle Y_g' n_H\}^{-1} \text{ s} . \qquad (8.8)$$

The curves in Fig. 8.2 are obtained from a study of cloud-cloud collisions in which the sputtering of the grain was followed through the lifetime of

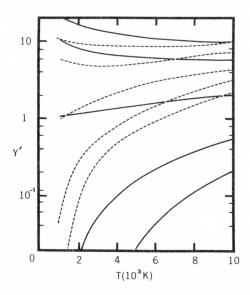

FIG. 8.2.   Sputtering yields as a function of gas temperature for a series of shock velocities (1, 5, 10, 15, and 18 km s^{-1} proceeding upwards). The solid curves are for He on H_2O; the broken curves are for H on CH_4.

the shock. We can approximate the results as follows. For a relative velocity of 30 km s^{-1} between clouds the shock velocity is 15 km s^{-1}. The
grain is exposed to a gas with $T_g \simeq 6000$ K, $n_H \simeq 10$ cm^{-3}, and relative
velocity 4 km s^{-1} for a time of about 500 y. Under these conditions $t_{sp} \simeq$
400 y for He on H_2O, and so an icy mantle is effectively sputtered away as
shown in the detailed calculations. Only the mantles are sputtered, however; refractory grains and cores of core-mantle particles would remain.
If the shock velocity is reduced even the mantle sputtering is less effective, because both the temperature and density in the shocked region
decrease. Thus 30 km s^{-1} is the critical cloud-collision velocity required
to destroy grain mantles. (A somewhat higher velocity would be necessary
if the calibration of the sputtering yields was high, as recent experimental data have suggested (see Fig. 11.4).) Note that sputtering is only
important in these transient events; longer exposure to the hot intercloud
phase of H I is offset by lower density and temperature.

Another interesting application of sputtering is to grain-mantle destruction in H II regions. In an H II region a grain probably has a positive charge as a result of photoemission. Therefore the values of Y'_g given
above would not be applicable because they are for neutral particles. The
sputtering yields from protons and helium ions suffer from two effects, the
decrease in impact cross-section and the decrease in kinetic energy of ions
that do strike the surface. Reduction by a factor of 100 or more can be
expected. If we take $T_g \simeq 10^4$ K, $n_H \simeq 10$ cm^{-3}, and $Y'_g \simeq 10^{-3}$ we find $t_{sp} \simeq$
10^8 y. In more dense H II regions it is possible that sputtering could be
an important mechanism over the lifetime of the region, but over the
shorter time ($\sim 10^4$ y) in which radiation pressure on grains contributes
most to the dynamical evolution, the grains are not likely affected. Optical erosion (photodesorption) may have a more substantial effect.

For application to higher temperature shocks, where refractory grains
like graphite and silicates can be sputtered, an empirical approach to
sputtering yields has been used. Experimental data are used to provide
information on how thresholds and sputtering yields depend on various parameters. These expressions are then numerically integrated to find the
average yield in a given environment. Fig. 8.3 shows values of nt_{sp} as a
function of T_g. Also labelled along the top is the velocity of an adiabatic strong shock required to produce T_g. In most of the calculations the
grains were taken to be at rest relative to the gas because the slowing-
down time (t_{gh}) was smaller than the radiative cooling time of the gas
(t_c). The requirement that $t_{sp} < t_c$ gives the following critical shock
velocities: 100 km s^{-1} for 'ice', 200 km s^{-1} for silicates, and 300 km s^{-1}
for iron and graphite. Nuclei of He are the most effective sputtering
agents.

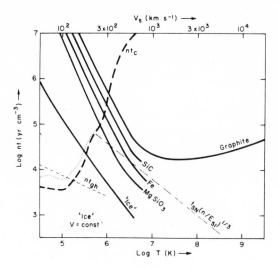

FIG. 8.3. Variation of nt_{sp} with T (lower abscissa) and V_s (upper abscissa) for several materials. The curves labelled 'ice' were calculated for a sublimation energy of 0.1 eV and a grain radius a=0.1 μ. For all other materials a=0.03 μ. The curve labelled V=const refers to ice grains moving at constant velocity V_s relative to gas at temperature T; all other sputtering curves refer to grains at rest. The effect of grain charge was included. Also shown are the cooling times (labelled nt_c) for a gas mixture of cosmic abundances, both for equilibrium cooling (broken curve) and for time-dependent cooling from a temperature of 10^6 K (dotted curve). The lifetime of a spherical adiabatic blast wave propagating through a uniform gas of density n/4 is labelled $t_{sn}(n/E_{51})^{1/3}$, where E_{51} is the initial blast wave energy in units of 10^{51} ergs. The slowing-down time scale of uncharged grains by encounters with gas atoms is labelled nt_{gh}. (From M.J. Barlow and J. Silk, Sputtering in interstellar shocks: a model for heavy element depletion, Astrophys. J. Lett. 211, L83 (1977). University of Chicago Press. ©1977 by the American Astronomical Society. All rights reserved.)

If the initial velocity of the blast wave were much higher, say 10^4 km s^{-1} from a supernova explosion, then the grain may not even survive until the shock has adiabatically cooled and slowed down to the 'critical velocity' just mentioned. Radiative cooling in a fast, hot shock is negligible compared with the adiabatic cooling. The latter is determined by how fast the supernova remnant expands, and is characterized by the time required to double the current supernova mass by sweeping up more inter-

stellar material (t_{sn}). Graphite will be sputtered away before the critical velocity is reached only if the ambient density $n/4 > 10$ cm^{-3}. Most of
the sputtering occurs at a relatively low velocity, $\sim$ 500 km s^{-1}. Silicates are more susceptible and disappear if $n/4 > 0.1$ cm^{-3}, which is a more
probable situation.

The severe depletion of gas-phase heavy elements commonly seen in the
interstellar medium (§10.1.2) is less pronounced for 'high velocity gas',
gas which has a velocity of about 100 km s^{-1} relative to normal material. A
consistent explanation can be developed as follows. Depletion is supposed
to be caused by condensation of the heavy elements in grains. Occasionally
a region is overrun by a powerful supernova shock, which accelerates and
heats the gas and sputters away the grains. Depletion is thereby eliminated in this material. Subsequently the supernova remnant sweeps up more
interstellar gas and dust at velocities below the critical velocity for
sputtering, resulting in intermediate values of depletion in gas at
intermediate velocities. Qualitatively that is what is observed.

In a quite different context, in hot gas in clusters of galaxies, the
sputtering rate is also of interest. X-ray observations show that $T_g \simeq 10^8$
K and $n \simeq 10^{-3}$ cm^{-3}. From Fig. 8.3 it can be seen that even the most stable grains, graphite, are sputtered in $t_{sp} \simeq 2 \times 10^7$ y. Thus while intergalactic grains might exist in clusters because of continuous injection
from galaxies, the equilibrium concentration will not be very large
(§12.3).

Electrostatic potential. The potential of a dust grain results from
the competition between three processes: photoelectric emission, sticking
of electrons, and sticking of positive ions. Each process is in turn
affected by the grain potential. As a result of Coulomb forces, the collision cross-section between a particle of charge Z (Z=-1 for an electron)
and a grain of potential U becomes $\sigma_d(1-2ZeU/mv^2)$, where m is the particle
mass and v is the initial relative velocity. The rate averaged over a Maxwellian distribution is then $\varepsilon\sigma_d\langle v\rangle$, where

$$\varepsilon = \exp(-ZeU/kT_g) \qquad\qquad (8.9)$$

when ZU>0 or

$$\varepsilon = 1-ZeU/kT_g \qquad\qquad (8.10)$$

when ZU<0. A positive grain potential also reduces the probability of photoemission, since lower-energy photoelectrons have insufficient energy to
escape. A self-consistent solution incorporating all of these effects must

be sought.

Consider first the balance between the sticking of positive ions and electrons in the absence of photoemission. If the grain were neutral the number of impacts by electrons would outnumber those by positive ions, because the thermal velocity of the electrons is so much higher. Given comparable sticking probabilities for ions and electrons, the grain would therefore become negatively charged. This gain is self-limiting because the rates are affected by the potential (eqns (8.9) and (8.10)). The equilibrium potential, which corresponds to an equality in sticking rates, is therefore determined from

$$m_e^{-\frac{1}{2}} S_e \exp(eU_s/kT_g) = m_i^{-\frac{1}{2}} S_i f_i (1 - Z_i eU_s/kT_g) \tag{8.11}$$

where f_i is the fraction of the total positive charge (which is equal to the number of electrons) carried by ion species i, and the right-hand side is summed over i. For simplicity assume that protons dominate in H II regions and that C II dominates in H I regions (ionization chiefly by starlight); then the solutions of eqn (8.11) for $S_i \approx S_e$ are

$$eU_s = -2.5 \ kT_g \quad (H \ II), \quad \text{and} \quad eU_s = -3.5 \ kT_g \ (H \ I). \tag{8.12}$$

The potential is therefore typically 2.2 V in an H II region where $T_g \approx 10^4$ K, and 100 times lower in an H I region.

The corresponding charge of a spherical dust particle in an H II region is

$$Z_d = -2.5 \ akT_g/e^2 = -150(a/10^{-5} \ cm)(T_g/10^4 \ K). \tag{8.13}$$

However, for typical grains in H I regions the mean excess of electrons from this process is very low and fluctuations around the mean would be comparable with the mean. Such a low charge is likely to obtain only in dense H I clouds where the internal extinction is sufficient to block out ultraviolet photons which might otherwise produce photoemission.

Inclusion of the photoelectric effect is straightforward in principle. Eqn (8.14), which balances photoemission against the sticking of electrons (positive ions are unimportant if the potential is positive), shows what parameters must be known:

$$n_e S_e \ \sigma_d \langle v_e \rangle \ (1 + eU_p/kT_g) = \int N_\nu \ yQ_a \sigma_d \ d\nu \approx N \ yQ_a \sigma_d \tag{8.14}$$

where $N_\nu d\nu$ is the ultraviolet photon flux (total N) and y is the yield (the probability of ejection of a photoelectron on absorption of a photon).

Note that absolute values of the parameters must now be known. Initial
estimates which suggested that photoemission in H I regions would be negligible have been subsequently revised to the point where photoemission could
be the dominant effect.

The photon spectrum of interest in H I regions ranges from 13.6 eV
down to a threshold energy of about 10 eV where y becomes small (y=0 when
the photon energy falls below the work function, somewhat less than 10 eV
for most materials; we use 10.1 eV in Fig. 8.4). From models based on number densities and spectra of early-type stars, and a few direct measurements of the local ultraviolet flux, $N=N_o \simeq 3 \times 10^7$ cm^{-2} s^{-1} over this
energy interval. For normal-sized interstellar grains $Q_a \simeq 1$, while for
smaller grains Q_a may be significantly less. (This could keep the average
albedo up (§10.3.2).) We shall let n_e and T_g be free parameters.

The other parameters S_e and y are determined by laboratory measurements. Limited experiments for electron impacts on neutral surfaces give
$S_e \simeq 0.3$-1 in the relevant energy range (eU+kT $\simeq$ eU $\simeq$ a few eV). On actual
interstellar grain surfaces it seems reasonable to assume $S_e \simeq 1$, since the
potential is significantly attractive and a variety of surface imperfections should enhance the sticking. The yield is somewhat more uncertain.
The highest yields for normal incidence on plane surfaces are ∿0.1-0.2 for

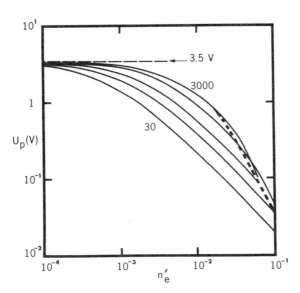

FIG. 8.4. Photoinduced electric potential of a grain in a gas
with normalized electron density n_e' (cm^{-3}) and temperature as
labelled (30, 100, 300, 1000, and 3000 K). The broken curve
gives the negative of the potential reached in the absence of
photoemission (independent of n_e').

good insulators (silicates and Al_2O_3), falling to ∿0.03 for semiconductors (SiC, graphite), and $<10^{-2}$ for metals (Fe). Ices such as H_2O have work functions comparable with 13.6 eV and so the yield is negligible. The yield is thus a strong function of composition. The yield also depends on the depth in the particle at which the photon is absorbed, and so the yield for small grains is different than for the plane-parallel situation encountered in the laboratory experiments. Some computations for small spheres, using internal fields from the Mie theory to calculate the divergence of the Poynting vector, indicate an enhancement by a factor of 2 or 3. Surface imperfections may also increase the yield. All of these values, however, apply to neutral surfaces, for which a typical photoelectron energy is ∿2 eV. Thus the yield will decrease as the grain charges up. This effect can be accounted for approximately by increasing the threshold from 10 to (10+eU) eV, so that $N < N_o$.

In Fig. 8.4 the value of U_p(eV) is plotted against $n_e' = n_e S_e Q_a^{-1}$(3 × $10^6/N_o y$) for several values of T_g. With the above estimates of the various parameters $n_e = n_e'$. Note the saturation in potential for low n_e' imposed by the cut-off in the photon spectrum at 13.6 eV and the quenching of the positive potential at large n_e', when $n_e' > 5 \ T_g^{-\frac{1}{2}}$. Inside a dense H I cloud the photon flux N could be much lower than the value N_o suggested for unshielded regions. Quenching will occur at a proportionately lower value of n_e. The locus of values of n_e' at which $U_p = -U_s$ is also given for comparison.

Photoemission can be expected to be even more prominent in the inner parts of H II regions where the ultraviolet flux is high. Furthermore the presence of photons in the Lyman continuum would mean that the positive value at which U_p saturates could be much higher than the value 3.5 V in H I regions. Fig. 8.5 shows some representative results. Notice the dominance of the photoeffect close to the star and the asymptotic approach to U_s in the outer region where the flux is diluted. These results may be scaled to other values of n_e simply by multiplying the abscissa by $(n_e/100$ $cm^{-3})^{1/6}$ (and truncating at 1.0). In an H II region with a higher density a larger fraction of the volume will contain positively charged grains. This result may be understood as follows. Let r_o be the radius at which U=0; clearly $r_o \propto n_e^{-\frac{1}{2}}$. However, this is more than offset by the behaviour of the Stromgren radius since $r_s \propto n_e^{-2/3}$. Thus $r_o/r_s \propto n_e^{1/6}$.

It seems quite certain that most grains will possess a significant potential most of the time. There are three important dynamical effects to be discussed: strong coupling of the translational motion of the grain to the magnetic field, increased drag from the plasma through which the grain may be moving, and precession of the angular momentum vector of the grain around the interstellar magnetic field.

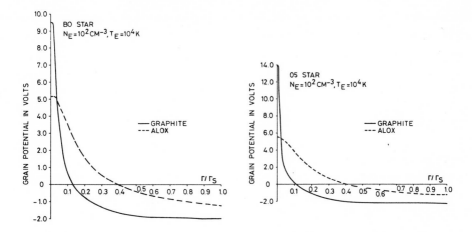

FIG. 8.5. Grain potential as a function of position in an H II region. (From A.F.M. Moorwood and B. Feuerbacher, Grain charging in H II regions, in Solid state astrophysics (eds. N.C. Wickramasinghe and D.J. Morgan), Reidel, Dordrecht (1976).)

8.2. TIME CONSTANTS OF GRAIN MOTION

The translational and rotational motions of a grain are affected by a variety of interactions with other components in the interstellar medium. Each interaction can be characterized by a time constant t; intercomparison of the time constants establishes the relative importance of the various effects. Numerical examples evaluated with a consistent set of parameters are summarized in Table 8.2. Values are in units of 10^{12} s. To scale these numbers reference must be made to the formulae given later.

Magnetic effects. Let us begin with how the motion of a grain with charge $Z_d e$ is affected by the presence of a magnetic field of strength B. If the grain has a velocity component transverse to the field, it will gyrate around the field direction at the cyclotron angular velocity ω_B. A suitable time constant for this motion is

TABLE 8.2
TIME CONSTANTS FOR GRAIN MOTION

t_B	t_L	t_p	t_ℓ	t_m	$t_{\ell i}$	t_i	t_a
0.45	0.54	190	5.0	6.7	0.5	0.008	4.0

Time constants in units of 10^{12} s calculated for $a = 10^{-5}$ cm, $s_d = 1.2$ g cm^{-3}, $B = 10^{-6}$ G, $U = 10^{-2}$ V, $n_H = 10$ cm^{-3}, $T_g = 10^2$ K, $K = 2.5 \times 10^{-12}/T_d$ s, $T_d = 10$ K, $n_i = 10$ cm^{-3} and $T_i = 10^4$ K.

$$t_B = \omega_B^{-1} = m_d c / Z_d e \ B \rightarrow 300 \ m_d c / aUB \ . \tag{8.15}$$

The arrow is used to indicate an expression for a spherical grain, here with potential U V. Using $B = 10^{-6}$ G, $U = 10^{-2}$ V (surely both underestimates), $a = 10^{-5}$ cm, and $s_d = 1.2$ g cm^{-3} we find $t_B = 4.5 \times 10^{11}$ s. This estimate, which can be considered to be an upper limit, is the first entry in Table 8.2.

We note that despite the high mass-to-charge ratio the gyro radius $r_B = t_B v_t$ of the grain motion is quite small. Even if the transverse velocity v_t is taken to be 10 km s^{-1}, typical of macroscopic ('cloud') motions in the interstellar medium, $r_B \approx 0.1$ pc, which is small compared with 'cloud' dimensions. Thus large-scale translational motion of the grain can occur only along the field direction.

The rotation of a charged grain is also affected by the field because the motion of the surface charge distribution generates a magnetic moment (parallel to the angular momentum vector). The torque produced by the interaction of this moment with the magnetic field results in precession of the angular momentum vector around the field direction at the Larmor frequency ω_L. The time constant is

$$t_L = \omega_L^{-1} = 2 \ Ic/<a^2>Z_d eB \rightarrow 1.2 \ m_d c/Z_d eB = 1.2 \ t_B \tag{8.16}$$

where I is the moment of inertia and $<a^2>$ is the mean square distance of the charges from the rotation axis. The expressions on the right, valid for a spherical grain with a symmetrical distribution of charge, emphasize the close relationship of t_L and t_B.

The interstellar magnetic field also induces a magnetic moment in a grain. When the grain is rotating about a direction transverse to the field the induced moment, which is not established instantaneously, produces a dissipative torque equal to $\chi_2 VB^2$, where χ_2 is the imaginary part of the (volume) magnetic susceptibility. This magnetic relaxation of the transverse components of the angular momentum is the basis for the Davis-Greenstein grain-alignment mechanism. Since grain angular velocities ($\omega \approx 10^5$ s^{-1}) are typically much lower than the characteristic dispersion frequency for magnetic relaxation ω_o, we can write $\chi_2 = K\omega$ where the parameter K depends on the type of magnetism. Both paramagnetism of impurity ions (as proposed originally) and nuclear paramagnetism of hydrogen in ices give $K \approx 2.5 \times 10^{-12}/T_d$ s. Combining these expressions we find

$$t_p = I/KVB^2 \rightarrow 0.4 \ a^2 s_d /KB^2 \tag{8.17}$$

the damping time for transverse angular momentum.

According to the Barnett effect a rotating body becomes magnetized as if it were placed in an external field $B_b = (2m_e c/eg)\omega \simeq 10^{-7}\omega$ G where $g \simeq 1$ is the gyromagnetic coefficient. The magnetic moment induced in the grain produces precession around the interstellar magnetic field with time scale t_b. By noting that the real and imaginary parts of the susceptibility are approximately related by $\chi_2/\chi_1 \simeq \omega/\omega_o$ when $\omega \ll \omega_o$, we find a convenient estimate $t_b/t_p \simeq \chi_2 B/\chi_1 B_b \simeq 10^{-4}\omega/\omega_o$. We also note that $t_b < t_L$.

Linear motion. The continual bombardment of a grain by gas particles tends to damp out the linear (and angular) momentum of a grain. We shall evaluate the frictional effect for the usual case in which the grain velocity is insignificant compared with the thermal motion of the gas particles. The time constant is the e-folding time for the momentum. For either of two types of impact, specular reflection or sticking followed by uncorrelated evaporation, the damping time for linear motion is

$$t_\ell = 3\, m_d/{<}v{>}n_g m_g A = 0.75\, t_m \rightarrow as_d/{<}v{>}n_g m_g \qquad (8.18)$$

where A is the surface area of the grain (the orientation of the grain with respect to its motion is assumed to be random), t_m is the time in which the grain is struck by its own mass in gas, and the factor ${<}v{>}n_g m_g$ is to be summed over all species in the gas. The entry 5×10^{12} s in Table 8.2 is for a pure hydrogen gas with $T_g = 100$ K (${<}v{>} = 1.45 \times 10^5$ cm s^{-1}) and $n_H = 10$ cm^{-3}. Incidentally, the time $(n_H {<}v{>}\sigma_d)^{-1}$ ⊖etween successive collisions with a hydrogen atom is only 2×10^3 s. Comparison of t_ℓ and t_B re-emphasizes the constraint on the motion produced by the magnetic field.

When the speed u of the grain is much in excess of the thermal speed of the gas particles (u > ${<}v{>}$) then the drag on the grain is dependent on the amount of material swept up along its path. In terms of previous time constants, $t \simeq t_m {<}v{>}/u$; note how it depends now on the speed of the grain.

If the grain is charged and moving through a partially ionized medium, the drag from the accumulated effects of long-range electrostatic encounters can substantially exceed the drag from actual impacts. The appropriate time constant is

$$t_i = 3\, m_d (2\pi)^{\frac{1}{2}} (kT_i)^{3/2}/8\pi\, m_i^{\frac{1}{2}} n_i Z_i^2 Z_d^2 e^4 \ell n\Lambda$$
$$\rightarrow 2(kT_i/Z_i eU)^2\, t_{\ell i}/\ell n\Lambda \qquad (8.19)$$

where $\ell n\Lambda = (3/2Z_i Z_d e^3)(k^3 T_i^3/m_e)^{\frac{1}{2}}$. The parameter Λ accounts for the diminished influence of long-range encounters past a certain distance, and $t_{\ell i}$ is the time constant from impacts of the same ions treated as neutral particles. Numerical values are insensitive to $\ell n\Lambda$, which is 20 ± 3 for a

wide range of possible interstellar conditions. Using eqn (8.12), with the
constant -2.5,

$$t_\ell/t_i = (t_\ell/t_{\ell i})(t_{\ell i}/t_i) = 60 \; n_i/n_g \; (m_i/m_g)^{\frac{1}{2}}(U/U_s)^2 \; . \tag{8.20}$$

In an H II region the electrostatic drag clearly dominates. Values appropriate to $T_i = 10^4$ K and $U = U_s$ are entered in Table 8.2. In an H I region,
however, there are a number of possibilities. If $U = U_s$, $t_\ell < t_i$ even for
substantial fractional ionization ($n_i/n_H \simeq 10^{-2}$, $m_i = m_H$). On the other hand
if photoemission is as important as estimated above, then for example,
$t_\ell/t_i \simeq 10^4$ when $n_i/n_H \simeq 5 \times 10^{-4}$, $n_H < 10$ cm^{-3}, and $m_i/m_H \simeq 12$, and t_ℓ/t_i
$\simeq 1$ when $n_i/n_H \simeq 10^{-2}$, $n_H > 10$ cm^{-3}, and $m_i = m_H$. The decreased effectiveness of ion encounters with increasing fractional ionization is caused by
quenching of the photoinduced potential.

    Angular motion. The damping time for angular motion depends somewhat
on the axis around which a non-spherical grain is rotating. With the same
assumptions about sticking as above we have

$$t_a = 0.6 \; f_a t_m + 0.6 \; t_m = 0.8 \; t_\ell \tag{8.21}$$

where the factor f_a ($0.5 < f_a < 2$) allows for non-spherical grains. It is
important that, as for t_B/t_ℓ, $t_L/t_a < 1$. Recall also that $t_b < t_L$. Therefore, through precession of the angular momentum vector, the magnetic field
becomes the axis of symmetry of the angular momentum distribution of the
grain. This ensures that any alignment of interstellar grains is with respect to the Galactic magnetic field.
    The relative importance of magnetic dissipation can be measured by

$$\delta = t_a/t_p + 2 \; KB^2/a\langle v\rangle \; n_g m_g \; . \tag{8.22}$$

Qualitatively $\delta > 1$ would be required for significant magnetic alignment.
Reference to Table 8.2 indicates that either K or B would have to be much
higher than we have assumed (§8.4). The situation is aggravated when
proper account is taken of the non-spherical shape.

8.3. RADIATION PRESSURE

The rate at which a plane wave of intensity I carries linear momentum
across a unit area normal to the direction of propagation is I/c; this can
be called the radiation pressure of the directed wave. A grain which
intercepts a portion of the wave, either by absorption or scattering, will

feel a net force, written as $C_r I/c$. This defines C_r, the cross-section for radiation pressure, as the effective area over which the pressure is exerted. Consideration of conservation of momentum shows that

$$C_r = C_s \ (1 - \langle \cos \Phi \rangle) + C_a \ . \tag{8.23}$$

The first term originates in radiation scattered out of the incident direction through a scattering angle Φ; the average is therefore weighted by the phase function p. The second term comes from absorption; there is no contribution from the re-emitted radiation, for which the angular distribution is assumed to be isotropic, or at least symmetrical. Note that C_r will in general depend on the angle of incidence and the state of polarization of the incoming wave through the dependences of C_s, C_a, and p which have been suppressed in eqn (8.23). Similarly $\langle \cos \Phi \rangle$ is orientation and polarization dependent; recall, however, that for a spherical particle such dependence vanishes and the average is called g (eqn (4.11)).

An efficiency factor Q_r is sometimes defined also. The behaviour of Q_r in various limits follows from the previous discussions of how Q_s, Q_a, and g vary. In the long-wavelength limit $Q_r = Q_a = Q_e$. The relationships for large spheres are summarized in Table 4.2. Usually $Q_r < Q_e$, because $Q_r = Q_e - \langle \cos \Phi \rangle Q_s$. The dependence of Q_r on size and wavelength is illustrated in Fig. 8.6 for spherical particles with fixed refractive index (cf. Figs. 2.3, 4.3, and 4.4).

One interesting application is to the motion of a grain subject to

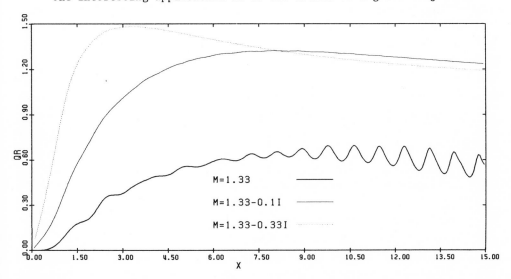

FIG. 8.6. Efficiency factor for radiation pressure (see Figs. 2.3, 4.3, and 4.4).

radiation pressure and gravitation near a star. For a given particle the
ratio of the radiative to gravitational acceleration is independent of the
distance to the star as long as the line of sight to the star is optically
thin; however, this ratio is a function of particle size. The ratio of
forces is

$$\beta = L_* \bar{C}_r / 4\pi cGM_* m_d \to 0.58 \, s_d^{-1} \, (M_\odot/M_*) \, (L_*/L_\odot) \, (\bar{Q}_r/10^4 a) \qquad (8.24)$$

where $\bar{Q}_r$ is averaged over the spectrum (e.g. the Planck mean). The size
dependence thus enters as $\bar{Q}_r/a$. For absorbing particles $\bar{Q}_r/a$ is largest
for sizes smaller than the wavelength ($x < 1$), but for purely scattering
particles $\bar{Q}_r/a$ peaks at intermediate values of x.

Whenever $\beta > 1$, grains can be driven away from the star. Setting $\beta = 1$
in eqn (8.24) therefore defines a critical luminosity L_d. Collisional
transfer of momentum imparts a net outwards force to the ambient gas as
well. Because the same radiative force is spread over a greater effective
mass, the critical luminosity L_{gd} for removal of gas and dust is larger, in
proportion to the mass of gas relative to dust. In the intermediate case
in which $L_d < L_* < L_{gd}$, gas remains bound to the star, but the grains still
drift outwards, their motion impeded but not stopped by frictional drag
from the gas.

Having estimated both the radiative force and the drag on the grain
from the ambient gas (time constant τ), we can write the equation of motion
(neglecting gravity) for a spherical grain

$$du/dt = L_* \bar{C}_r / 4\pi R^2 cm_d - u/\tau . \qquad (8.25)$$

There are two limiting cases. If the drag is insufficient, then the second
term is never important and the grain accelerates to

$$u_{max}(R) = \{(3L_* \bar{Q}_r / 8\pi cs_d a) \, (R_i^{-1} - R^{-1})\}^{\frac{1}{2}}$$

$$\to 4.7\times10^2 \, \{s_d^{-1} (L_*/L_\odot) \, (R_\odot/R_i) \, (\bar{Q}_r/10^4 a)\}^{\frac{1}{2}} \; km \; s^{-1} . \qquad (8.26)$$

(If the grains drag along a lot of gas, eqn (8.26) can still be used
approximately by including the increased effective mass in the factor s_d.)

The drag from the ambient gas severely limits the velocity in the
other limiting case obtained in eqn (8.25) by setting $du/dt=0$. This
defines the drift velocity as

$$u_{dr} = \tau L_* \sigma_d \bar{Q}_r / 4\pi R^2 cm_d . \qquad (8.27)$$

The choice of formula for τ depends on the ionization of the gas (and potential of the grain) and on whether u_{dr} is subsonic or supersonic.

Several examples will be given to illustrate the relevance of these results in a variety of astronomical situations. Application of eqn (8.24) to the solar system leads to an interesting result (§9.5): submicron-sized particles can be expelled, whereas larger (millimetre-sized) grains are retained. We note, however, that the latter do not have stable orbits because their orbital motion is effectively damped through the Poynting-Robertson effect. (Since aberration produces an excess of light falling on the leading edge of a grain the orbital (tangential) motion u_t is opposed by a force equal to u_t/c times the radial force from radiation pressure.) Large grains therefore spiral slowly in towards the sun, until they evaporate completely or become small enough to be blown back out (§9.5). A value $u_{max} \simeq 50$ km s^{-1} for a small grain starting in the inner solar system is possible.

Circumstellar grains near late-type stars represent another important application. Consider a small graphite particle (a $\simeq 0.05$ μ) near a luminous cool star ($T_* \simeq 2500$ K, $L_* \simeq 10^4$ $L_\odot$, $R_* \simeq 535$ $R_\odot$, $M_* \simeq 4$ $M_\odot$). Since $\bar{Q}_r/a \simeq 3.5 \times 10^4$ for this grain size and stellar temperature, the ratio of forces is very large, $\sim 2.2 \times 10^3$. Note the possibility for expulsion of the circumstellar gas as well, since $L_* \gg L_d$. The composition of the particle is also quite important; a silicate particle of the same size has much less absorption ($\bar{Q}_r/a \simeq 0.06 \times 10^4$) so that $\beta \approx 40$.

Graphite grains might form near the surface of a cool carbon star, perhaps as close as $R_i \simeq 1.2$ $R_* \simeq 640$ $R_\odot$ in our example; therefore $u_{max} < 2.3 \times 10^3$ km s^{-1}. It is unlikely that such a high velocity will be realized, because the necessary high luminosity probably results in $L_* > L_{gd}$ and hence frictional drag from the gas swept along with the dust. (Conversely, if $L_* < L_{gd}$ the dust can separate from the gas, but the value of u_{max} corresponding to this L_* is lower.) Guided by the high value of u_{max} we use τ appropriate to supersonic motion (§8.2) to find

$$u_{dr}^2 = L_* \bar{Q}_r / 4\pi R^2 c \rho_g \rightarrow L_* \bar{Q}_r v_f / c \dot{M} . \qquad (8.28)$$

To obtain the right-hand formulation we have assumed a steady mass loss model, characterized by flow velocity v_f and mass loss rate $\dot{M}$. Substitution of $v_f \simeq 20$ km s^{-1} and $\dot{M} \simeq 10^{-6}$ $M_\odot$ y^{-1} gives $u_{dr} \simeq 25$ km s^{-1}. This is a significant reduction from u_{max}, but is still comparable with v_f and large enough for the effects of sputtering to be considered. The origin of the outflowing material from which grains condense is another problem. Possibly grains can form in the photosphere during cool stages of a pulsation cycle or in cool regions of convective-cell structure and thus drive

the mass loss. A self-consistent picture for this example results if
$(\rho_g/\rho_d)(R_i/R_*) \simeq 10^4$. All of the above numerical values are of course for
illustration only. However, more detailed modelling of individual stars,
where observations warrant it, would be based on the same principles.

A third example is an ice grain ($a \simeq 0.3$ μ) in an H II region, illu-
minated by an O5 star ($T_* \simeq 5 \times 10^4$ K, $L_* \simeq 3 \times 10^5$ $L_\odot$, $M_* \simeq 40$ $M_\odot$). Here
$\bar{Q}_r/a \simeq 3.3 \times 10^4$, and so the ratio of forces is again very high, $\sim 1.2 \times$
10^4. If acceleration were to begin within 2 pc of two O5 stars (ice grains
closer in than this would melt), then $u_{max} \simeq 60$ km s^{-1}. Frictional drag
from the gas is dominated by electrostactic interactions and reduces u_{dr} to
subsonic motion ($<v> \simeq 14$ km s^{-1}). Using the formula for t_i, $n_H = 10$ cm^{-3},
and $|U| = U_s$, we find $u_{dr} = 0.7$ km s^{-1}.

The radiative force transferred to the gas in this example is also
substantial. In fact time-dependent models show that the central region of
such an H II region is swept clear of dust and gas on a timescale of 2 $\times$
10^5 y. During this time the drift of the grains relative to the gas is
negligible. However, in regions with later-type central stars the separation of grains from gas may be significant during the main-sequence lifetime. If radiation pressure on grains is an important factor then some
observed H II regions should have central holes. The Rosette Nebula is a
prime candidate for exploring such a mechanism in detail; W3(A) may contain
a less evolved example.

8.4. GRAIN ALIGNMENT

An interstellar grain is subject to a variety of random fluctuating torques
so that even if it is initially at rest it is soon set rotating. The
increase in angular velocity is a diffusion process which is ultimately
limited by frictional torques. A Fokker-Planck equation can be used to
describe the time development of the angular momentum distribution of an
ensemble of grains. For illustration we consider the simple case of spherical grains exposed to isotropic disturbing forces. Writing the distribution as $f(J)J^2 dJ$ we have

$$\partial f/\partial t = J^{-2} \partial\{J^2(-Rf + \tfrac{1}{2}\partial Df/\partial J)\}/\partial J \tag{8.29}$$

which in equilibrium gives

$$f(J) \propto \exp[\int_0^J \{2 R'(J')/D(J')\} dJ'] \tag{8.30}$$

where $R'=R-\tfrac{1}{2}\partial D/\partial J$. For the processes described below the diffusion coefficient D, which by symmetry is one-third the mean square change in angular

momentum, has no momentum dependence. The frictional term R is simply J/τ, where τ is a time constant for angular motion, such as described in §8.2. This leads to a Boltzmann distribution characterized by a temperature

$$T = \tau D/2Ik \; . \tag{8.31}$$

When several processes are competing, individual values of R and D are summed so that

$$\tau = (\Sigma \; \tau_j^{-1})^{-1} \; , \quad \text{and} \quad T = \Sigma \; D_j / \Sigma \; (D_j/T_j) = \Sigma \; T_j R_j / \Sigma \; R_j \; . \tag{8.32}$$

To be specific consider bombardment by hydrogen atoms, a major contributor to the disalignment of interstellar grains. If the atoms stick and subsequently evaporate at the grain temperature then

$$D_a = (4\pi/3) \; \langle v \rangle \; n_g m_g T_a \tag{8.33}$$

where $T_a = \frac{1}{2}(T_g + T_d)$. Combining this with t_a (eqn (8.21)) we see that the equilibrium rotational temperature from gas bombardment is T_a. Next consider rotation about a direction transverse to the magnetic field. We have seen how to calculate the damping time t_p for paramagnetic absorption (eqn (8.17)). The equilibrium rotational temperature due to the magnetic interaction is T_d, since fluctuations of the internal magnetic moments depend on the grain temperature. We could calculate D_p, but instead proceed to combine the gas-bombardment and magnetic processes to find from eqns (8.22) and (8.32)

$$T_{av} = (T_a + \delta \; T_d)/(1+\delta) \; . \tag{8.34}$$

Rotation around the magnetic field direction is still characterized by T_a. This illustrates, by a particular important case, a general feature of what is required for any grain-alignment mechanism: a difference in rotational temperatures for different directions in space. Such a difference breaks the symmetry in the angular momentum distribution. The alignment of the angular momentum vector in this example is with respect to the magnetic field. In any case precession of the angular momentum vector makes the field an axis of symmetry for the distribution, since precession proceeds on a time scale much faster than any other process.

As we have seen above (§3.2.2) the quantity q_{SB} is a useful measure of the efficiency of any alignment process. Generally q_{SB} cannot be predicted readily, but in the case of magnetic alignment Monte Carlo simulation of the processes involved has pointed the way to a useful approximation.

Consider first two special cases for which there are analytical results. In a Cartesian co-ordinate system the angular momentum distribution of a spherical particle is the product of three Boltzmann factors, characterized by temperatures T_{av}, T_{av}, and T_a. From this the distribution of $\underline{J}$ with respect to $\underline{B}$ can be written down; with some further algebra the desired expression (cf. eqn (3.11)) is

$$q'_{JB} = q_o(T_{av}/T_a-1) = q_o\{\delta(T_d-T_g)/(T_d+T_g)(1+\delta)\} . \tag{8.35}$$

The function $q_o(x)$ is actually an analytical expression, but is plotted for convenience in Fig. 8.7. The Monte Carlo results indicate that eqn (8.35) is suitable for axially symmetric particles if δ is the value appropriate to rotation about an axis perpendicular to both the symmetry axis and the magnetic field. The corresponding correction factors which have been worked out for eqn (8.22) will not be given here. Note the importance of the relative size of T_d and T_g; in particular when $T_d=T_g=T_{av}$ there is no alignment (q_{JB} changes sign).

     The second analytical result concerns the distribution of angular momentum for an axially symmetric particle subjected to only isotropic bombardment by the gas. The distribution is again the product of three Boltzmann factors, now with the same temperatures but different moments of inertia, say I and γI, the latter for rotation about each of the two transverse axes. Similarly it may be shown that

$$q'_{SJ} = q_o(\gamma-1) . \tag{8.36}$$

The Monte Carlo results show this approximation to be an underestimate of the actual value when $T_d/T_g<1$ (as expected in the interstellar medium) and

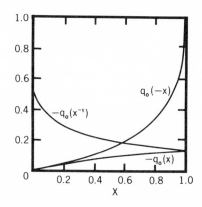

FIG. 8.7.   Graphical representation of the alignment function q_o.

an overestimate otherwise. This arises because J is aligned with respect
to B, so that the temperatures for the three Boltzmann factors are not
really the same.

Finally we write, following eqn (3.12),

$$q_{SB} = f\, q_{SJ}\, q_{JB} = f'\, q'_{SJ}\, q'_{JB}\,. \tag{8.37}$$

The factor f, found to be slightly greater than unity, allows for the correlation of the angles (S,J) and (J,B). Of interest for computations using
the analytical results is the factor f'. Generally f' lies in the range
1-3; it is largest when the degree of alignment is small, and always
$|f' q'_{SJ}| < 1$.

To illustrate how weak the alignment might be in the interstellar
medium we present one numerical example with the following parameters: a
prolate spheroid with axial ratio 2:1 ($\gamma = 2.5$), $T_d/T_g = 0.11$, and $\delta = 4$.   From
our equations $q'_{JB} = 0.21$ and $q'_{SJ} = -0.17$; from the Monte Carlo results
$q_{JB} = 0.21$, $q_{SJ} = -0.28$, and $q_{SB} = -0.080$; therefore f=1.3, f'=2.2, and
$|f' q'_{SJ}| = 0.37$.   If this grain has a radius 10^{-5} cm, the magnetic field
required to produce $\delta = 4$ in a gas of density 10^{-23} g cm^{-3} with $T_g = 100$ K is
1.5×10^{-5} G.   Such a field strength is an order of magnitude larger than
that which exists.

A concise method of summarizing many more Monte Carlo results is Fig.
8.8, where values of $|q_{SB}|$ for three different grain shapes are displayed
as a function of B. Fig. 8.8 can in turn be used with eqn (3.15) to predict $(p/\tau)_V$ as a function of B. We have done this using computed values of
$(p/\tau)_{pf}$ at λ_c and choosing the grain size such that $\lambda_c \approx \lambda_V$ as is observed.
The figure caption explains how the magnetic field scales with volume; a
slight shift is required to switch from rectangular prisms to spheroids.
Thus the original curves are displaced both vertically and horizontally to
produce Fig. 8.9 for the same three shapes and three refractive indices.
Also indicated is the observed maximum $(p/\tau)_V$ which is to be explained.
Clearly magnetic alignment is far from adequate if B is a few microgauss
and K derives from paramagnetism.

One solution to this difficulty could be that the grains possess special magnetic properties. The whole grain might be ferromagnetic (e.g.
magnetite), in which case there is magnetic ordering within one or more
domains. More plausible is the view that paramagnetic ions within a non-
magnetic host grain may be clustered sufficiently to act collectively as
single magnetically ordered domains (superparamagnetism). The relaxation
is thereby enhanced without changing the overall concentration of magnetic
impurities from a low value of a few per cent. In each case K would exceed
the value adopted above by several orders of magnitude. Recalling that the

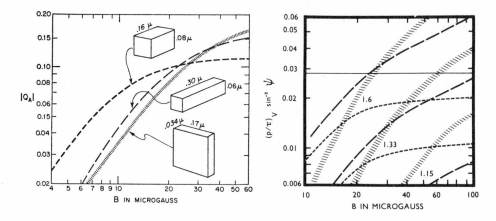

FIG. 8.8. Axial alignment measure $Q_A = q_{SB}$ for three prism-shaped grains of approximately equal volume $V_d = 10^{-15}$ cm^3, as a function of magnetic field. Assumed are $T_g = 100$ K, $T_d = 11$ K, and $K = 2.5 \times 10^{-12}/T_d$ s. For similar grains of different size, B $\propto V_d^{1/6}$ for the same Q_A if other parameters including T_d are held constant. If the dependence of T_d on grain size is allowed for, however, B $\propto V_d^{0.12}$ approximately. If T_d were changed while keeping V_d constant, then B $\propto T_d^{0.9}$ for the same Q_A. (From P.A. Aannestad and E.M. Purcell, Interstellar grains, A. Rev. Astron. Astrophys. 11, 309 (1973). Reproduced with permission. ©1973 by Annual Reviews Inc.)

FIG. 8.9. Predicted values of $(p/\tau)_V$ using the results of Fig. 8.8, but scaling the size so that $\lambda_c = \lambda_V$. Extinction cross-sections for spheroids rather than prisms have been used. The interstellar maximum is indicated by the thin horizontal line at 0.028.

field required drops as $K^{-\frac{1}{2}}$, we see that magnetic alignment in the weak interstellar field would be possible.

That there is a limit to the degree of alignment as the field or K increases is indicated by Fig. 8.8. If we use the limiting values of $|q_{SB}|$ (see Fig. 3.6), we can calculate allowed combinations of m and the axial ratio that would give the observed value of p/τ. Fig. 3.7 shows the result of this exercise for prolate and oblate spheroids, along with the similar, but weaker, limits established above for any alignment process. More stringent limits would apply if the polarizing grains do not account for all of the extinction.

Another possible modification relates to the value of q_{SJ}. If there is efficient dissipation of rotational kinetic energy by internal friction,

as we shall describe later, then the grains spin preferentially around the
axis with the largest moment of inertia. In that case q_{SJ} would approach
its limiting value (-0.5 or 1), $|q_{SB}|$ would be correspondingly larger, and
the limits on size and refractive index shown for Davis–Greenstein alignment in Fig. 3.7 could be relaxed somewhat.

There have been a number of other proposals as to how the required
asymmetry in rotational temperatures might develop. Prominent among these
are supersonic streaming of grains relative to the gas, and exchange of
angular momentum with photons. The former would lead to alignment opposite
to Davis–Greenstein alignment because charged grains can stream only along
the direction of the magnetic field. Sustaining the streaming velocities
over large regions in space seems quite unlikely, but localized applications might be found. The latter depends mainly on the intrinsic angular
momentum of absorbed photons which could have an effect since the illumination of a grain is principally from the direction of the Galactic plane.
However, in conditions usually encountered in the interstellar medium the
quantitative contribution to alignment is negligible. A more significant
contribution is made to disalignment, since the number of photons emitted
by a low-temperature grain greatly exceeds the number absorbed. As pointed
out above, precession of the grain in the magnetic field would inevitably
smear out any alignment but that with respect to the magnetic field. However, no proposal in which precession could play a role has withstood
detailed examination.

Interstellar grains as pinwheels. The specific problems with Davis–
Greenstein alignment have led to re-examination of the rotation of interstellar grains. In the processes described above the rotational velocity
builds up in a diffusive manner, resulting in 'Brownian rotation' characterized by a rotational temperature similar to the gas kinetic temperature.
On the other hand, if there were a torque on the grain that was steady
(with respect to orientation), rather than random, the grain could be
accelerated systematically to high angular velocity. Again rotational drag
from the interstellar gas would limit the speed, but now at a rotational
temperature much higher than in thermal equilibrium. Magnetic relaxation,
opposing rotation transverse to the field, would proceed very slowly, typi-
fied as before by t_p; however, eventually the angular momentum vector of
the grain would end up perfectly aligned along the magnetic field, even
though the field is weak.

Given this conceptual change it has not been difficult to devise
mechanisms by which steady torques ('rockets') might arise; on the contrary
it may be difficult to dismiss the importance of this 'pinwheeling'. In
particular, slight variations (1 in 10^3) in the thermal accomodation coef-

ficient for H atoms or in the photoelectric emissivity over the grain surface and recombination plus ejection of H_2 molecules at favoured active
sites are attractive possibilities, producing rotational energies 10^3-10^6
times greater than thermal excitation.

Many details of this novel scheme have yet to be worked out, however.
One crucial question is that of the permanence of the 'rocket' site over
the time required for significant spin-up ($<t_a$) and alignment (t_p). Conceivably another time scale would have to be introduced: the time required
for the grain to grow or exchange a surface monolayer. This is much
shorter than $t_a \approx t_m$ by a factor of 10^{-2} or 10^{-3} and so the steady torques
would have a much shorter time over which to act. A stochastic treatment
would be required. It appears possible that individual grains would spend
sufficiently long periods of time ($>t_p$) in suprathermal motion about a single axis even when the 'rockets' reverse on a time scale much less than t_a
and that the fraction of grains poorly aligned at any one time is small.

Of equal importance is the dynamical problem of which grain axis is
preferred for rapid rotation. Here the question of internal dissipation is
of interest, for if the problem is simply posed as one in which a grain
subject to no external torques loses rotational kinetic energy then it is
clear that the grain must ultimately end up spinning around the axis with
the greatest moment of inertia. The source of dissipation might be elastic
or magnetic in origin; both involve the rapid precession of the angular
velocity vector around the symmetry axis of a non-spherical grain at a rate
$\Omega \approx (\gamma^{-1}-1)\omega$. We now estimate the time scales.

If the energies of rotation and elastic oscillation are denoted E_r
and $E_{e\ell}$ respectively then $E_{e\ell} \approx E_r \sigma$, where σ is the strain. However, $\sigma \approx$
$s_d^2 \omega^2 a^2/\mu$, the numerator being an estimate of the stress and μ being the
elastic modulus. The rate of dissipation due to imperfections in the ordered structure of the grains can be parameterized in terms of the quality
factor Q as $\dot{E} = E_{e\ell}\Omega/Q$, and so the time scale for dissipation is

$$t_{ed} = fQ\mu/s_d\omega^3 a^2 \rightarrow 10^{10} \text{ s .} \qquad (8.38)$$

Proper accounting for the geometrical factors shows that $f \approx 10$ independently
of γ (though there would be no effect for a spherical grain). We have
evaluated t_{ed} for a cubical grain with side $2a = 2 \times 10^{-5}$ cm, $s_d = 1.2$ g
cm^{-3}, $T_g = 100$ K (Table 8.2), $\mu = 10^{11}$ dyn cm^{-2}, and $Q = 10^4$, the latter
being suggested by limited theoretical and experimental work on cold materials at acoustic frequencies. We note that $t_{ed} < t_a$ and that the size
dependence ($a^{11/2}$) is considerable.

Magnetic dissipation occurs as the angular velocity vector moves in
body co-ordinates because there is a component of the Barnett field which

lags behind the instantaneous field. The time scale t_{md} is of the form of
eqn (8.17) for t_p for Davis-Greenstein relaxation with B replaced by B_b.
Since $(B_b/B)^2 \simeq 10^8$, $t_{md} \simeq 10^6$ s for parameters like those in Table 8.2.
More rigourous calculation for a rectangular brick shows that

$$t_{md} = (\gamma^3/|1-\gamma|)\{2a^2 s_d/3K\omega^2(mc/e)^2\} \tag{8.39}$$

where ω corresponds to rotation about the axis with the largest moment of
inertia and a^2 is the mean square radius about this axis. Note the impor-
tant dependence on $|1-\gamma|$ which emphasizes that there is no dissipation for
a spherical grain. For an oblate square brick with $2a = 2 \times 10^{-5}$ cm, a:b =
2:1, and other parameters as above we find $\gamma=5/8$, $\omega \simeq 3 \times 10^5$ s^{-1}, and t_{md}
$\simeq 2 \times 10^5$ s. The size dependence (a^7 for fixed γ) is more extreme than
that for t_{ed}. Nevertheless we conclude that both magnetic and elastic dissipation are sufficiently rapid that q_{SJ} will reach its maximum in a time
much less than t_a and the time for 'rocket' spin-up to about kT_g.

All of the ingredients for grain alignment are possibly in hand. In
summary, internal dissipation ensures that a grain spins around the axis
with the largest moment of inertia. 'Rocket' torques systematically build
up the rotation about this axis to suprathermal velocities and maintain
this orientation in body co-ordinates for a time exceeding t_p. Magnetic
relaxation in the weak interstellar magnetic field has sufficient time to
align the angular momentum vector along the field and the resulting electric vector orientation is parallel to the magnetic field as required by
observations. The attendant high degree of alignment places the least
stringent restrictions on the axial ratio and refractive index of the
grains (Fig. 3.7). A detailed treatment combining these components is
awaited with interest.

FURTHER READING

§8.1

LEE, T.J. (1976). The influence of grain mantles on the formation of
    hydrogen molecules on grain surfaces. In Solid state astrophysics
    (eds. N.C. Wickramasinghe and D.J. Morgan). Reidel, Dordrecht.

WATSON, W.D., and SALPETER, E.E. (1972). Molecule formation on interstel-
    lar grains. Astrophys. J. 174, 321.

SPITZER, L., and JENKINS, E.B. (1975). Ultraviolet studies of the inter-
    stellar gas. A. Rev. Astron. Astrophys. 13, 133.

AANNESTAD, P.A. (1973). Molecule formation. II. In interstellar shock
    waves. Astrophys. J. Suppl. 25, 223.

BARLOW, M.J. (1978). The destruction and growth of dust grains in interstellar space. I. Destruction by sputtering. Mon. Not. R. astron. Soc. 183, 367.

----. (1978). The destruction and growth of dust grains in interstellar space. II. Destruction by grain surface reactions, grain-grain collisions and photodesorption. Mon. Not. R. astron. Soc. 183, 397.

ONAKA, T., and KAMIJO, F. (1978). Destruction of interstellar grains by sputtering. Astron. Astrophys. 64, 53.

SPITZER, L. (1968). Diffuse matter in space. Wiley, New York.

WATSON, W.D. (1972). Heating of interstellar H I clouds by ultraviolet photoelectron emission from grains. Astrophys. J. 176, 103.

DRAINE, B. (1978). Photoelectric heating of interstellar gas. Astrophys. J. Suppl. 36, 595.

§8.2
See §8.1. (Spitzer 1968).

MARTIN, P.G. (1971). On interstellar grain alignment by a magnetic field. Mon. Not. R. astron. Soc. 153, 279.

JONES, R.V., and SPITZER, L. (1967). Magnetic alignment of interstellar grains. Astrophys. J. 147, 943.

DOLGINOV, A.Z., and MYTROPHANOV, I.G. (1976). Orientation of cosmic dust grains. Astrophys. Space Sci. 43, 291.

§8.3
See §2.2.1. (Wickramasinghe 1973).

SALPETER, E.E. (1974). Formation and flow of dust grains in cool stellar atmospheres. Astrophys. J. 193, 585.

MATHEWS, W.G. (1967). Dynamic effects of radiation pressure in H II regions. Astrophys. J. 147, 965.

§8.4
PURCELL, E.M., and SPITZER, L. (1971). Orientation of rotating grains. Astrophys. J. 167, 31.

PURCELL, E.M. (1975). Interstellar grains as pinwheels. In The dusty universe (eds. G.B. Field and A.G.W. Cameron). Neale Watson Academic Publications, New York.

9

DUST IN OUR GALAXY

9.1. THE LARGE-SCALE DISTRIBUTION

Interstellar dust is concentrated in the plane of our Galaxy, as are H I gas and population I stars. Perhaps the most striking evidence of this concentration is the appearance of dark obscuring lanes in the Milky Way. These lanes or rifts, prominent enough to be seen with the unaided eye, are seen in considerable detail in Fig. 9.1, a composite photograph of part of the Galactic plane. The patchy distribution of material has led to the concept of clouds, on a variety of length scales, to which we shall return (§9.2). Other spiral galaxies viewed edge-on show a similar concentration of dust to the mid-plane (Fig. 12.1). In galaxies that are viewed more face-on dust can be seen to trace out the spiral arms (Fig. 12.2).

We have seen that $E'_{B-V} \simeq 0.6$ kpc^{-1} or $A'_V \simeq 2$ kpc^{-1} in the Galactic plane within a few kiloparsecs of the sun. If this average were to apply throughout the plane, the Galactic centre would suffer 20 mag of optical extinction. While this value is uncertain, it is sufficient to illustrate why the Galactic centre region can be studied only at infrared and longer wavelengths, and why in general we have little optical knowledge of the dust in the far reaches of our Galaxy. At low galactic latitudes few galaxies can be seen either, leading to the term 'zone of avoidance'.

FIG. 9.1. Composite view of the Milky Way from Scorpius (right) to Cassiopeia (left). Prominent objects, from which the scale can be judged, are the ρ Oph complex (upper right, see Fig. 9.4, brightest star α Sco = Antares), α Aql = Altair (lower centre), α Lyr = Vega, and M 31 = the Andromeda Galaxy (lower left). (Hale Observatories photograph.)

The <u>cosecant law</u>. The layer of dust is quite thin. Reddening measurements of stars as a function of distance and galactic latitude show the
dust usually lies within 50 to 100 pc of the Galactic plane. Maps of far-
infrared thermal emission should help delineate the distribution at larger
distances from the sun (§7.1). If this thin layer is assumed to be uniform, a crude approximation, then the latitude dependence of the reddening
of objects outside the layer arises from the pathlength variation and may
be written

$$E_{B-V} = E^o_{B-V} \ \text{cosec}|b| \ . \qquad\qquad (9.1)$$

Here E^o_{B-V} is the inferred reddening towards the Galactic poles, derived
from the slope of the cosecant law at lower latitudes. Independent fits to
colour data for RR Lyrae stars, globular clusters, spiral galaxies, and
brightest members of galaxy clusters are consistent with E^o_{B-V}=0.06. This
is compatible with $E'_{B-V} \simeq 0.6 \ \text{kpc}^{-1}$ (derived from b≈0°) and a layer semi-
thickness of about 100 pc. On the other hand, numerous photoelectric
investigations of the colours of distant field stars near the Galactic
poles show virtually no reddening; in fact $E_{B-V} < 0.01$ for $|b| > 50°$,
except for a few small regions of sky. The explanation of this discrepancy
is a puzzle. Galaxy counts, mapping of neutral hydrogen column densities
and optical polarization in the polar caps, and high-latitude reflection
nebulosity (§5.1) all indicate that the distribution of interstellar material is patchy. Scatter about the mean cosecant laws shows this as well.
Possibly the field stars sampled just happen to lie in areas of abnormally
low reddening. The direct measurements might also be made consistent with
the value E^o_{B-V} from the cosecant law by supposing that the sun lies in a
local minimum in the dust density in the Galactic plane. In particular, if
the sun is near the axis of a cylindrical hole with radius about the layer
semi-thickness (radius ≈ 50-100 pc), then there is little reddening above
$|b| \simeq 50°$ but a cosecant dependence at lower latitudes; such behaviour is
compatible with the available reddening data. Further direct evidence for
this simple model is provided by reddening measurements of stars within 200
pc of the sun and 40 pc of the plane; the local reddening is also very
patchy and extending out to about 50 pc, $E^o_{B-V} < 0.01$.

   A local minimum is less surprising when considered in the context of
the large-scale distribution of interstellar reddening in the Galactic
plane. One representation, based on colour excesses and distances (assuming R=3.0) for thousands of stars, is shown in Fig. 9.2. Many cloud complexes can be seen, separated by relatively clear areas. The cloud complexes themselves must actually consist of many smaller clouds, according
to the scatter about the mean reddening-distance relations. Much of the

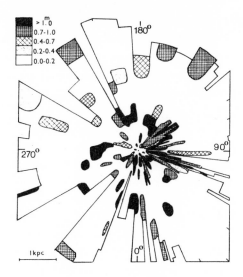

FIG. 9.2. Large-scale distribution of interstellar dust within
a few kpc in the Galactic plane. Shading indicates the change
in colour excess per kpc (E'_{B-V}). (After M.P. Fitzgerald, The
distribution of interstellar reddening material, Astron. J.
73, 983 (1968).)

dust appears to be associated with the spiral structure features delineated
by early-type stars and reflection nebulae.

Several attempts have been made to measure the optical thickness of
the Galactic absorbing layer using counts of galaxies to a given magnitude
limit. In each analysis a decrease of counts with increasing cosec $|b|$ is
seen, along with considerable scatter attributed to the cloudy distribution
of dust just mentioned. However, the slopes A_B^o found in the different
investigations are quite discrepant, ranging from 0.25 to 0.5. The most
recent discussions of the selection effects and calibration corrections
involved favour the lower value, $A_B^o \simeq 0.25$. However, the question probably
cannot be considered to be settled. Our interest in finding a definitive
value lies in the ratio A_B^o/E_{B-V}^o = R+1. Since this is the ratio of slopes
of similar functions of cosec $|b|$ it should not be affected seriously by a
possible hole in absorption towards the poles. A value of $A_B^o = 0.25$,
together with $E_{B-V}^o = 0.06$, leads to R=3.2, in excellent agreement with the
average value 3.3 from other determinations (§3.1). A higher value of A_B^o
would introduce the possibility of large particles and neutral extinction.
Neutral extinction caused by material spread out over distances more than
100 pc from the plane is difficult to detect by any other technique.

9.2. THE CONCEPT OF DUST CLOUDS

Having mentioned briefly the evidence for the patchy distribution of interstellar dust on a large scale, we turn to the methods of finding density
concentrations or 'clouds' on smaller scales. Dust clouds are sometimes
identifiable individually as reflection nebulae (§5.1) or as dark nebulae
and globules (§9.3). This specific information can be supplemented by statistical studies of the cloud structure, through analysis of the fluctuations in reddening and polarization from star to star in a limited area of
the sky. Systematic measurements of stars in an open cluster are often
used for these purposes, because the additional complication of varying
distance is removed. Some of the basic results will be discussed below.

    The correlation of gas and dust. The distribution of interstellar
gas has also been studied, mostly with radiofrequency line observations. A
central role is played by 21-cm emission line studies of H I; more recently
surveys of CO and other molecular species have received increasing attention. Interstellar gas is found to have a cloudy structure also, and so
the question of the correlation of dust with gas arises. In some cases the
close association of dust and gas is obvious. For example, large molecular
clouds are often discovered coincident with dense dark dust clouds (e.g.
the Coalsack); this is not surprising since dust not only shields the
molecules against destructive ultraviolet radiation, but also might cata-
lyse some molecule formation. The complex dust lanes that run through H II
regions provide another example from optical data alone (e.g. Fig. 9.6).
Detailed 21-cm line mapping of selected areas usually shows a correlation
of gas with photographed dark cloud structure. However, in some cases
involving dense clouds there is an anticorrelation, suggesting the conversion of much of the H I into H_2.
    More generally gas and dust abundances have been compared along many
different lines of sight in the Galaxy by a variety of techniques. Usually
the amount of dust is measured by the colour excess E_{B-V} of a particular
star or cluster. The amount of gas on the same line of sight can be
detected in several ways. The emission profile at 21-cm can be used, but
there is the problem of separating out the emission from gas more distant
than the star. One solution has been to use globular clusters, which lie
outside the H I layer of the Galaxy. There are still potential problems
with self-absorption of the line emission, and the failure to detect the
gas in H II and H_2 regions which can contribute to the measured E_{B-V}. The
true correlation of gas and dust is best revealed by plotting a scaled neu-
tral hydrogen column density $N_H \sin|b|$ against $E_{B-V} \sin|b|$ which removes
the spurious correlation arising from path length variations. On the basis

of several determinations the slope of the correlation line is about 5.0, in the units of eqn (9.2).

Lyman-α absorption also measures the H I column density; again H II regions are missed but the corrections can be estimated. However, since molecular hydrogen is detected by its ultraviolet absorption lines, it is possible to include it in the total column density. The results from initial satellite observations of bright stars within 1 kpc lie in the range 5.4-7.7. For each star the deviations of the hydrogen column density and E_{B-V} from the average amounts expected for that distance can be computed. The existence of these deviations indicates a cloudy distribution for interstellar material, and the good correlation found for the deviations confirms that dust and gas are closely associated. Note that the Lyman-α and 21-cm techniques give consistent results, even though quite different volumes of the Galactic disk were sampled.

Finally the total hydrogen density can be determined indirectly by X-ray absorption measurements. For the sources considered the X-ray absorption is detected at sufficiently high energies (>0.6 keV) that most of the absorption arises from He, C, N, O, and heavier elements, and so the state of hydrogen (H I, H II, or H_2) is unimportant; the relative abundances of elements with respect to hydrogen are needed, however. Only diffuse sources were used to avoid potential confusion with gas associated with some of the more compact sources. Two separate investigations find 6.6 and 6.8, again in close agreement with the stellar data, despite characteristically larger path lengths and absorption.

The average gas-to-dust ratio is

$$N_H(total)/E_{B-V} \simeq (6.0\pm1.5) \times 10^{21} \text{ cm}^{-2} \text{ mag}^{-1} \tag{9.2}$$

consistent with $E'_{B-V} \simeq 0.6 \text{ kpc}^{-1}$, $n_H(total) \simeq 1.2 \text{ cm}^{-3}$, and a gas-to-dust mass ratio of about 100 (§3.1.2).

Fluctuations of reddening. There are a number of indications that the interstellar dust distribution is lumpy on a fairly small scale. For example in the cosecant law determinations mentioned there is always considerable scatter about the mean relation. Furthermore the reddening and polarization from star to star varies noticeably, even for stars in nearly the same direction and at the same distance. The reddening and polarization data can be used to derive some average properties of the interstellar medium, such as the number of clouds along a line of sight and the optical depth per cloud. Even the characteristic size of a cloud can be determined if the correlation of reddening or polarization is measured as a function of angular separation of stars at the same distance.

A simplistic model involving only one type of cloud can be developed for illustration, but should not be taken literally. Suppose, as additional assumptions, that these clouds are distributed randomly in space, that each produces a reddening E_o, and that for stars within the distance interval being examined n clouds are intercepted along an average line of sight. Then E_o and n can be determined from the ensemble mean reddening <E> and the standard deviation σ_E of the reddening about the mean by using the relations

$$<E> = nE_o \quad \text{and} \quad \sigma_E^2 = <E>^2/n .$$
                                                                    (9.3)

We shall use the B-V colour system below.

Open clusters provide an opportunity to pursue this analysis, though the volume of space sampled is quite restricted. Stock 2 is a well-studied representative of the several clusters that have been examined. It lies at a distance of 316 pc towards $\ell=133°$, $b=-1.7°$, in a region of fairly uniform polarization and higher reddening than usual. Using <E>=0.375 and σ_E=0.12, a single-cloud model gives n≈10, ν≈31 (the number of clouds per kiloparsec), E_o≈0.04, and νE_o≈1.2.

This technique has also been applied at the other extreme to field stars satisfying |z|<100 pc and r<1 kpc. In these studies account had to be taken of the dispersive effects of the range in stellar distances within each distance subinterval examined. It was also found that by using a model with two cloud types (more free parameters!) the skewed shape in the distribution of reddening could be fitted. The results of one such determination are given in Table 9.1. It should be emphasized that since these results are based on an oversimplified model they should be taken only as an indication of the scale of structures in the interstellar medium detect-

TABLE 9.1
ILLUSTRATIVE VALUES OF CLOUD PARAMETERS

Parameter	Units	Values	
		Small Clouds	Large Clouds
E_o	mag	0.08	0.5
υ	kpc^1	5	0.5
υE_o	mag kpc^1	0.4	0.2
D	pc	4	70
E'_o	mag kpc^1	20	7
n_c	kpc^{-3}	3×10^5	1×10^2
f		0.02	0.04
N_H	cm^{-2}	5×10^{20}	3×10^{21}
n_H	cm^{-3}	40	15
fn_H	cm^{-3}	0.8	0.6
M	$M_\odot$	65	1×10^5

able by this technique, and should not be accepted literally as the basis
for other investigations.

With this inherent limitation clearly understood, we can take a
further step and use the angular correlation of the reddening to find the
linear dimension of the clouds, if a further assumption about their distribution along the line of sight is made (usually uniform). The data for
Stock 2 undoubtedly show the effects of correlation at small angular sepa-
rations and suggest a characteristic size (diameter) of 2 pc. The field-
star data do not give as reliable a value for the radius of a 'small' cloud
because of the lack of closely spaced stars; a diameter of about 4 pc is
indicated. On the other hand the 'large' cloud diameter, ∿70 pc, is somewhat better determined. These parameters can then be manipulated to find
other characteristics of the clouds, such as those given in Table 9.1 (f is
the filling factor). Again it would be a mistake to take these numbers as
being more than illustrative. Nevertheless the existence of small-scale
structure in the interstellar medium is uncontroversial.

Reflection nebulae and H I clouds. Reflection nebulae provide some
additional insight. The fact that the majority of bright candidate stars
do not illuminate reflection nebulae shows that the dust distribution is
non-uniform and that the filling factor for the more dense regions is low.
The well-observed Hubble relation between apparent size of the nebula and
apparent magnitude of the star (eqn (5.1), or linear size and absolute magnitude) can be understood if the nebula is 'illumination bounded' rather
than 'density bounded'. However, for a given magnitude there is a considerable scatter in size, by a factor of 5 or more. Some of this is no doubt
caused by the effects of nebular geometry, optical depth, and internal reddening. If actual density cut-offs cause the scatter, then the Hubble
relation is better taken as the envelope of the observed correlation and
significant density contrasts on a scale of 10^{-1}-10 pc are revealed.

The Pleiades nebulae (Fig. 5.1) may be taken as illustrative of
'density-bounded' nebulae. They are individually quite asymmetric with
respect to the illuminating stars (especially Merope); furthermore Pleione
and Taygeta have little nebulosity. Here we also have a good example of
hierarchical structure. Filaments within the nebulosity have sizes in the
range 10^{-2}-10^{-1} pc, whereas the nebulae measure about a parsec across. The
total system extends to at least 10 pc. The Pleiades cluster is moving
with a radial velocity of 10 km s^{-1} with respect to the intervening gas, as
revealed by interstellar Na I, Ca II, and CH^{+} optical absorption lines and
by 21-cm emission lines. The velocities of the nebulae are unfortunately
not measured, but it seems likely that the visible nebulae are the result
of the chance passage of the bright stars and relatively dense dust concen-

trations. Thus in these and similar reflection nebulae a rare glimpse of
the fine structure in the interstellar medium is obtained. (Some of the
fine structure may be induced during the encounter.)

At this point it is worthwhile making further reference to studies of
the structure in the interstellar gas distribution. Most detail is
obtained from spatial and velocity mapping, particularly with the 21-cm
emission line. While there is evidence for 'clouds' not unlike those in
Table 9.1, the overwhelming impression is one of the variety and hierarchical nature of the distribution. What might be called clouds are usually
ragged in outline (as are reflection nebulae and dark nebulae) and often
exist in large groupings or cloud complexes. Perhaps aggregate, with the
connotation of embedding, is a more apt label. Some aggregates extend for
more than 100 pc. The 'large clouds' mentioned above may in fact also be
aggregates. However, aggregates are far from being spherically symmetric.
There is evidence both for flattened sheets (that might be related to
supernova explosions) and for elongated filamentary complexes (that might
be related to the magnetic field along which they often lie). As more and
more observational data have been accumulated it has become increasingly
clear that the long-standing concept of a random spatial distribution of
spherical clouds (of several types or even a continuous mass spectrum)
should be abandoned for all but order-of-magnitude estimates if a realistic
description of the interstellar medium is desired.

Fluctuations of polarization. A simple cloud model can also be
developed for the interpretation of interstellar polarization observations,
subject of course to the same reservations. Let there be on average m
clouds along a line of sight, each producing the same degree of polarization p_o but at different orientations θ_i with respect to some direction of
symmetry ($<\theta_i>$)=0). Then using equations like (2.14) and (2.18) we have

$$<q> = m \ p_o \ <\cos 2\theta_i> , \qquad <u> = m \ p_o \ <\sin 2\theta_i> = 0 \qquad (9.4)$$

$$\sigma_q^2 = m \ p_o^2 \ <\cos^2 2\theta_i> , \qquad \sigma_u^2 = m \ p_o^2 \ <\sin^2 2\theta_i> . \qquad (9.5)$$

Over the face of Stock 2, the open cluster mentioned above, the
direction of symmetry is in the Galactic plane, and both the direction and
degree of polarization are relatively uniform. The observed values, $<q> \approx$
$<p> \approx 2.3 \times 10^{-2}$ and $\sigma_q \approx 7.6 \times 10^{-3}$, indicate that m≈9. While this is in
good agreement with n≈10 from extinction, it appears from other evidence
that the 'clouds' found by the two methods are not identical. If the
clouds were identical then one would expect to see a better correlation of
reddening and polarization than is actually observed (Fig. 9.3). Also the

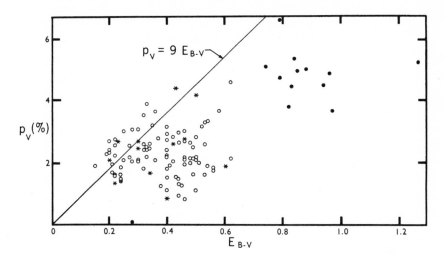

FIG. 9.3. Linear polarization against reddening for stars in
the region of cluster Stock 2. Cluster members are denoted by
open circles, probable members by asterisks, and non-members by
solid circles. The straight line indicates that the maximum
polarization-to-extinction ratio adopted above (eqn (3.10)) is
exceeded in this cluster. (From W. Krzeminski and K. Serkowski, Photometric and polarimetric observations of the nearby
strongly reddened open cluster Stock 2, Astrophys. J. 147, 988
(1967). University of Chicago Press. ⊕1967 by the University
of Chicago.)

angular correlation of the polarization of cluster members indicates a mic-
roscale, or cloud size, of about 0.3 pc, significantly smaller than for
extinction. The lack of correspondence should not be unexpected, however,
since there are many factors other than simply the amount of dust that
affect the polarization, in particular the degree of alignment and the
orientation of the magnetic field. The latter effect is illustrated by the
arching filaments in the Merope nebula (Fig. 5.1), along which the observed
electric vectors (and by deduction the magnetic field) lie; thus the polarization of stars viewed through different parts of a single cloud can be
quite different.

By using the ratio σ_u/σ_q we can make use of that part of the dispersion related to the different position angles of polarization produced by
each cloud. The ratio is 0.8 for Stock 2, so that the root mean square θ_1
is about 23°. This is a better indicator of the large amount of disorder
in alignment along the line of sight than is the dispersion in observed
position angle, $\sigma_\theta \simeq 8°$, because the latter depends on the net polarization
from many clouds. In fact within the framework of the simple cloud model

$\sigma_\theta \simeq (\theta_i)_{rms} m^{-\frac{1}{2}}$, a relation giving self-consistent results for Stock 2.
The remarkable implication is that even in this direction of seemingly good
alignment the magnetic field in localized volumes of space exhibits large
fluctuations in direction.

This point can be pursued further by examining σ_θ, σ_q, and σ_u as
functions of galactic longitude. The pronounced minima in σ_θ, which occur
at longitudes (135°, 310°) where the average polarization is a maximum and
parallel to the Galactic plane, have already been cited (§3.2.3) as evidence for a longitudinal component of the field running perpendicular to
this direction. When σ_q and σ_u are derived for stars within restricted
distance intervals, which reduces the dispersive effects related to distance, it is found that they are of comparable magnitude and that they show
little systematic dependence on longitude. Thus the minima in σ_θ are not
the result of smaller fluctuations of the field direction at those longitudes, but rather are the consequence of the large average polarization.
Along the mean field direction no such systematic polarization component
arises, and so we observe lower polarization and a large scatter in position angles produced by the superposition of randomly oriented polarizing
'clouds'.

The observed development of linear polarization with distance can
also be understood qualitatively if there are large fluctuations in field
direction. Over shorter distances, several hundred parsecs, polarization
increases linearly, but beyond 1 kpc there is considerable depolarization
from the superposition of disoriented elements and so the degree of polarization becomes saturated. It is possible to develop a statistical model to
analyse the polarization produced by a medium that is irregular on a variety of length scales. Such a treatment probably provides a more realistic
description of those effects of the patchy distribution of dust and its
uneven alignment that have been revealed by the simpler discrete-cloud
model described here.

Interstellar circular polarization, which depends on changes in the
position angle of the magnetic field along the line of sight, potentially
supplies information on the fluctuations revealed by the linear polarization data. Unless there is a systematic twist in field direction the average circular polarization in any longitude range should be zero, but the
dispersion σ_v is non-vanishing. In a medium in which the fluctuating component of the field is comparable with the mean, σ_v should depend little on
longitude. On the other hand, since $\sigma_G \simeq \sigma_\theta$ (see eqn (2.24) for G), σ_G
should have the same minima, for the same reasons. While preliminary
observations tend to confirm this schematic picture, a much more detailed
study is warranted.

9.3. DARK CLOUDS

Dark clouds are relatively dense regions of interstellar dust that are discovered in silhouette against a luminous background provided by either a
rich star field or the diffuse Hα emission of an H II region. The dark
lanes of obscuration seen in the Milky Way and photographed in other spiral
galaxies are the largest scale of this phenomenon. The discovery of clouds
on a smaller scale using star counts obviously favours more dense and/or
nearby clouds (few foreground stars) for which the contrast is greatest.
The clouds of smallest angular extent are found only against a diffuse
background, because the statistical accuracy of star counts would be low.
The most recent of several catalogues of dark nebulae is based on the red
and blue photographs of the Palomar Observatory sky survey and has about
1800 entries. Examination of the distribution of these nebulae on the sky
shows most to be condensations within extensive cloud complexes.

    The concentration of dust clouds in the Galactic plane is reflected

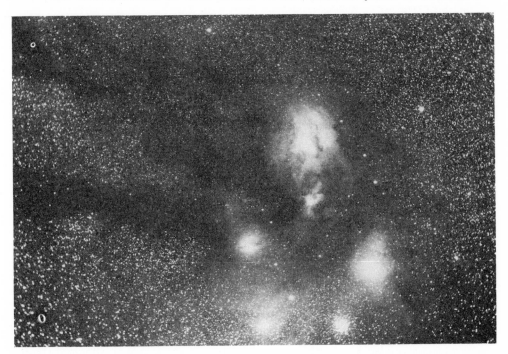

FIG. 9.4. Dark lanes and nebulosity in the ρ Oph complex. The
star at the centre is HD 147889. From the upper centre, counter-clockwise around the pentagon (N through E), are ρ Oph, 22
Sco, α Sco, M 4, and σ Sco. Another globular cluster M 80 is
at the upper right. Scale: 1 = ───────. (Yerkes Observatory photograph.)

in the latitude distribution. For most clouds $|b| < 10°$. Among the notable
exceptions are the Orion complex at $\ell \approx 210°$, $b \approx -20°$, and the ρ Oph complex
at $\ell \approx 350°$, $b \approx 17°$ (see Figs. 9.1 and 9.4). The latter clouds, at distances
of about 200 pc, are then 60 pc out of the plane, within the usual z range
of 100 pc. On the other hand the Orion clouds, some 400 pc away, are unusually far from the plane, about 140 pc. The longitude distribution is far
from uniform, there being far more dark clouds seen towards the Galactic
centre than the anti-centre. It is interesting to note that the direction
of most uniform polarization, $\ell \approx 140°$, is relatively free of dark clouds,
whereas regions of large disorder in position angle, like $\ell \approx 80°$, coincide
with major cloud complexes, in this case the Great Rift in Cygnus.

The large variety of shapes and sizes of dark clouds precludes a comprehensive description here. However, some impression of the complexity
can be gained by examining the selection of photographs given (Figs.
9.4-9.7). Of necessity these pictures are somewhat 'out of context',
though the scale is given in each case. To develop a better feeling for
the relationship of the clouds to their surroundings, it is rewarding to
look at all of these regions on sky survey photographs. We shall note a
few of the interesting properties.

Fig. 9.4 shows part of the dark cloud complex near ρ Oph. Particularly interesting are the dark lanes which converge in a cloud bounded by ρ
Oph, 22 Sco, and σ Sco. Towards the centre is the highly obscured star HD
147889 and its surrounding reflection nebulosity. The fine filamentary
structure is a characteristic of other regions as well (e.g. Fig. 9.5).

Elongation is a common property on all scales. However, there are
some clouds for which the strongest extinction is confined to a nearly circular region. Fig. 9.5 shows two globules of the type catalogued by Barnard; note the complete lack of stars as compared with other less opaque
regions that are seen blocking out diffuse Hα emission. Globules such as
these commonly have ragged edges, but it is not unusual to find one boundary rather sharp, as in B 92. This must arise from compression of the gas
cloud in which the dust is embedded.

Sharp edges are more easily seen against a bright H II region, such
as M 16 (= NGC 6611 = the Eagle Nebula) in Fig. 9.6. Larger contorted
structures have been called 'elephant trunks'; small opaque globules of
many sizes are also seen. Some of the dark clouds are outlined by bright
rims of Hα emission. This results as the ionization front penetrates into
the dense cloud material.

The smallest clouds or globules are detectable in silhouette against
the H II regions with which they are often associated. While not every H
II region shows dark globules, it is possible that there are unseen internal globules. In the Orion nebula, for example, it has been proposed that

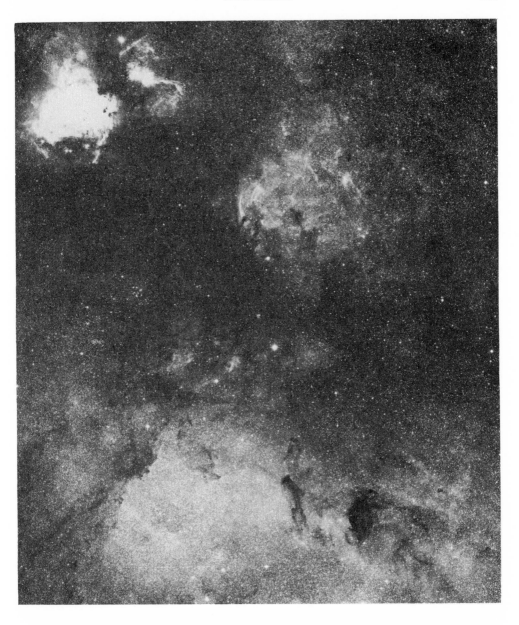

FIG. 9.5. Red-light photograph of region of Milky Way centred near $\ell=14°$, $b=-0°5$. North up, east to left; the Galactic plane runs roughly SSW to NNE. The prominent H II region is M 17. Seen against diffuse Hα emission and star cloud to the southwest are two Barnard dark clouds, B 92 (larger, on W) and B 93 (with ragged tail). Scale: 20 arc min = ——————. (©The National Geographic Society - Palomar Observatory Sky Survey. Reproduced by permission from the Hale Observatories.)

ionization fronts penetrating dense neutral globules are responsible for
the peculiar velocity fields. Because such globules are intrinsically
small they are also difficult to detect against a star field, despite their
opacity, unless the field is unusually rich and the cloud is close enough
to subtend an appreciable size.

  Physical characteristics. Many of the important physical properties
like density and mass depend on the size of the cloud, and hence on a determination of the distance. Of great historical importance, and still
widely used, are several related methods based on star counts on photographic plates. Automated microdensitometers now make this work much less
tedious. Basically, the dependence of the number of stars counted on
apparent magnitude is used to find two apparent magnitudes, that at which
the excess obscuration sets in and that at which it levels off, and also
the total amount of extinction. Clearly the luminosity function and space
distribution of stars must be understood in order to convert the data for
the two magnitude limits into distance and radial extent. In practice star
counts over an assumed uniform region of the cloud are compared with counts
in an apparently clear region (a Wolf diagram). General foreground extinction is assumed to affect each region equally. The rest of the analysis
rests on the assumption of a common space distribution and luminosity function in the two regions. The actual luminosity function must be known as
well, so that its dispersive effects on the count distribution (and hence
on radial extent) can be removed (e.g. Malmquist's method). Alternatively,
stars within a narrow range of spectral type have been used; the large number of spectra needed are obtained on objective prism photographs.
    Photoelectric photometry of stars with known spectral classification
is used to map out the largest clouds. We have seen this applied to delineate the large-scale distribution of dust (Fig. 9.2). There is another
technique (Becker's method), used in individual clouds, that is based
solely on photoelectric colour measurements (no spectra). Measurements for
stars of the same apparent magnitude are plotted in a colour-colour diagram, (U-V) versus (B-V). If there were no reddening the standard main
sequence locus would be seen; extinction shifts stars up along parallel
reddening lines. At a given apparent magnitude early-type stars are much
more distant, so that if the apparent magnitude chosen is faint enough the
early-type stars will be behind the cloud, while the late-type stars are in
the foreground. The colour at which the discontinuity in the colour-colour
plot occurs, combined with the apparent magnitude, yields the distance.
The amount of reddening in the cloud can also be found.
    Since the distance to an emission nebula can be measured by spectro-
scopic parallax once the exciting star is identified, upper limits are

FIG. 9.6. The Eagle Nebula, M 16, is a bright H II region
excited by O stars. Note the small dark Bok globules and the
dominant 'elephant trunk' outlined with a bright rim. Many
other ionization fronts can be seen penetrating dark neutral
regions. West is up, north to the left. Scale: 2 arc min =
───────────. (Kitt Peak National Observatory 4-m photograph.)

obtained for the distances of many nebulae seen in silhouette. When a dark
cloud has a bright rim, as is the case for the Horsehead nebula (B 33) near
σ Ori or in M 16, then a definite association can be made. An example of a
chance projection is provided by the familiar North America Nebula in Cygnus. Actually it is only a part of a huge distant H II region cut off from
the other major component, the Pelican Nebula, by a ragged portion of the
foreground dust complex.

The total extinction τ in the cloud, usually measured by photographic
methods, together with a few assumptions which will become clear, underlies
estimates of the column density, density, and mass of the cloud. The following formula for the total mass of a cloud with surface area πR^2 (it need
not be circular) illustrates the dependence on various parameters:

$$M \simeq 100 \ (4as_d\tau/3Q_e)\pi R^2 + 60 \ \tau\theta^2 d^2 \ . \tag{9.6}$$

It seems reasonable to have assumed a gas-to-dust mass ratio of 100, close
to the limit allowed by cosmic abundances, since this limit is found to be
approached even in more diffuse clouds where accretion is less favoured.
For the numerical result on the right-hand side we have adopted the units
100 pc, degrees, and solar masses for d, θ, and M respectively, and have
assumed that the grains are not too different than seen in normal inter-
stellar extinction ($a \approx 0.2$ μ, $s_d \approx 1$ g cm^{-3}, $Q_e \approx 2$). Evidence for selective
extinction can be seen by comparing dark clouds on red and blue photo-
graphs; the larger opacity at blue wavelengths not only makes the cloud
darker, but also extends the apparent size.

Similarly the total gas density, expressed as the number density of
hydrogen molecules, the dominant species, is

$$n_{H_2} \simeq 50 \ (4as_d\tau/3Q_e)/2m_H R + 40 \ \tau/\theta d \ . \tag{9.7}$$

For simplicity it has been assumed that the cloud is approximately round,
so that its depth can be judged from its angular extent. Molecular line
observations provide an alternative technique for estimating the mass if
the molecular abundance relative to hydrogen can be calibrated. CO is present whenever the optical extinction exceeds about a magnitude; H_2CO is
also studied frequently. Sometimes the molecular hydrogen density can be
deduced from the requirements for collisional excitation of other molecules.

Keeping in mind the large variety in cloud shapes and sizes we can
now turn to Table 9.2, where some cloud characteristics are summarized for
the purposes of illustration. These characteristics should be taken as
order-of-magnitude estimates, rather than representative values for dis-

TABLE 9.2

PROPERTIES OF A SELECTION OF DARK CLOUDS

Object	Distance (pc)	Angular Radius (arc min)	Radius (pc)	τ_B	Mass $(M_\odot)$	n_{H_2} (cm^{-3})
M 16: Small globule	2000	0.025	0.015	$>10^*$	>0.04	$>5 \times 10^4$
Large globule		0.25	0.15	$>10^*$	>4	$>5 \times 10^3$
Thumb Print Nebula	400*	3	0.35	5	10	10^3
ρ Oph cloud: optical cloud	170	60	3.0	7	1000	2×10^2
IR cluster		15	0.74	20	220	2×10^3
molecular cloud and far-IR source		5	0.25	25	30	7×10^3

* estimated

crete classes of cloud.   A few further words of description for each seem
appropriate.

Many of the dark globules in M 16 (Fig. 9.6) have bright rims,
clearly associating them with the nebula.   Since the opacity used is only
an estimated lower limit, even the smallest structures contain more than a
stellar mass.   The properties of a cloud like the larger globule resemble
those of several of the more nearby clouds noted by Barnard (e.g. B 92 in
Fig. 9.5).   The long elephant trunk is much more massive.   Notice also that
the densities found are quite high.   This accounts for the contrast of the
bright (Hα) rims with the more diffuse H II region.

Another larger globule is the Thumb Print Nebula shown in Fig. 9.7.
It is a rather excellent example of an isolated 'bright dark nebula', being
illuminated by the general interstellar radiation field.   It has a much
higher opacity than the high-latitude reflection nebulae mentioned above
($\S5.1$).   Star counts in the outer parts of the cloud support an r^{-2} density
distribution.   Monte Carlo scattering calculations confirm that the
observed surface brightness profile is the result of a peaked distribution
(like r^{-2}), rather than a uniform-density configuration which would have
produced a flatter core with a sharper bright ring.   Three basic observations, the radial position of the maximum in surface brightness, the
brightness difference between the bright ring and the dark core, and the
brightness of the ring above the diffuse sky background, lead to a relatively unique determination of three other quantities, the central optical
depth, the albedo, and the phase function (or g), for each bandpass
observed.   In this case $\tau \simeq 16$ at the centre, and so we have used an average
of 5 for Table 9.2; naturally the central value of n_{H_2} is several times the
average.   The grains are strongly forward throwing ($g \gtrsim 0.7$) and have a high
albedo ($\tilde{\omega} \approx 0.7$); these values are consistent with less well determined
values from diffuse Galactic light (Fig. 10.7).   Comparison of observations
at red, visual, and blue wavelengths shows that the extinction is selec-

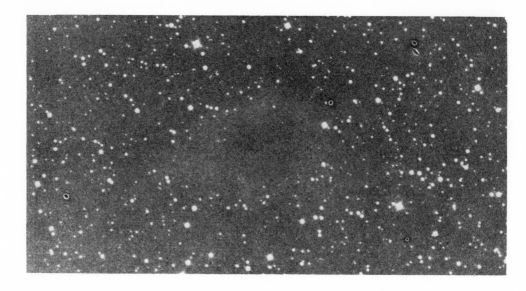

FIG. 9.7. The Thumb Print Nebula, a bright dark nebula in Chamaeleon. North up, east to left. Scale: 2 arc min = ————.
(Cerro-Tololo Inter-American Observatory Curtis-Schmidt photograph, courtesy of A.N. Witt.)

tive, but that the grains may be somewhat larger than normal. More observations of similar objects will be valuable.

The frequent association of dark clouds with molecular clouds and
regions of recent star formation and the r^{-2} density distribution of the
Thumb Print Nebula which is reminiscent of the characteristic central concentration built up in collapse models of interstellar clouds are both suggestive of the possibility that dark clouds are in a state of gravitational
collapse. For a preliminary investigation we can use the simple collapse
criterion

$$M > 2.4 \; kTR/\mu m_H G \; . \qquad\qquad (9.8)$$

The critical relation is $M=1.9TR$ where $\mu=2.4$, M is in solar masses, and R
is in parsecs. CO observations give T=10 K. This line is plotted in Fig.
9.8 for comparison with the values determined for a number of globules.
While the hypothesis of collapse looks promising it should be recalled that
several complications, such as magnetic fields, turbulent motions, and
rotation, have been ignored. These can be investigated with radio measurements of line widths and Zeeman splitting. Some preliminary observations
support collapse. This important question should be investigated for individual clouds as more observational data become available.

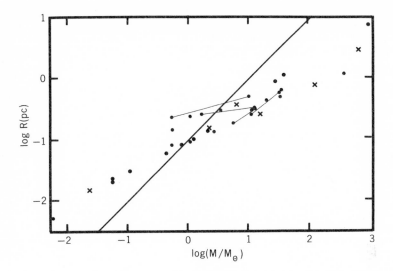

FIG. 9.8. Plot of radius against mass for representative dark
globules (thin lines joining different estimates for same globule indicate uncertainty). The crosses are for the data given
in Table 9.2. Globules to the right of the line are unstable
to gravitational collapse if T_g =10 K.

The ρ Oph cloud. The large cloud near HD 147889, south of ρ Oph at
the end of the dark lanes, has been the subject of detailed examination
because of its considerable mass and large angular scale (see Table 9.2).
Fig. 9.9 shows contours of visual extinction derived from star counts on
the Palomar Observatory Sky Survey photographs. There are a number of condensations apparent within the cloud; however, the contrast is limited by
the resolution and the maximum extinction $A_V \approx 6.3$ detectable by the star
counts. It can be seen in Fig. 9.8 that this cloud as a whole is probably
collapsing. The important discovery of 67 infrared sources (at 2 μ) has
demonstrated that some dense fragments of the cloud have already formed
stars. Infrared spectral distributions of these stars suggest they are the
brightest members (B3-F5) of a recently formed cluster, with a luminosity
function similar to other known clusters. Visual extinctions ranging from
10 to 30 mag are derived for these stars, confirming the high cloud densities ($n_{H_2} \approx 10^3$-10^5 cm^{-3}) implied by molecular line observations.
    In the same region as the cluster, radio continuum observations
reveal six compact H II regions, three actually coincident with 2-μ
sources. The strongest thermal source has a luminosity characteristic of a
relatively early B1-B2 main sequence star and has an extremely high density
$n_e \approx 4 \times 10^5$ cm^{-3}. The fact that it is not seen as a 2-μ source may be
explained by extreme obscuration, $A_V > 100$. Thus even near-infrared surveys

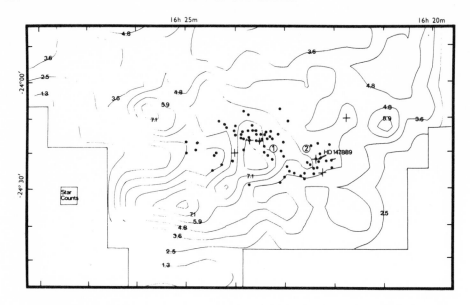

FIG. 9.9. Contour map of visual extinction near HD 147889 (see
Fig. 9.4) using star counts. The members of a cluster detected
at 2 μ are shown by dots. The continuum sources found at 11 cm
are indicated by pluses. Position 1 is the main peak in CO
emission and coincides with the far-infrared source; position 2
is an enhancement in CO emission arising from local heating,
possibly by HD 147889. (After P.J. Encrenaz, E. Falgarone, and
R. Lucas, CO, dust and H_2 in the molecular cloud near ρ Ophiuchi, Astron. Astrophys. 44, 73 (1975).)

of dark clouds may not penetrate to all the regions of active star formation.

A strong extended (5 arc min) far-infrared (350-μ) source is located
within the confines of the cluster, near some of the compact radio sources,
and coincident with the main peak (1 in Fig. 9.9) of CO emission. The dust
is probably heated by one or more massive stars which are forming in this
dense condensation. The similarities of this and the far-infrared source
in the Orion Molecular Cloud are striking.

As a molecular source in which CO, CS, OH, H_2CO, SO, and possibly NH_3
have been detected, this dark cloud, commonly called cloud 4, is quite
interesting. The molecule CO is present over a relatively extended region,
wherever $A_V > 2$. Its abundance, expressed as $N_{CO}/A_V \simeq 10^{17}$ cm^{-2}, together
with eqn (9.2), suggests that about 20 per cent of the available C is in
molecular form; the rest could be either in atomic form or condensed in
dust particles.

The kinetic temperature is 25-30 K, considerably higher than 10 K

which is typical of dark clouds. Gravitational collapse, for which we have
seen ample evidence, could provide the enhanced heat input. In addition
stellar and infrared sources apparently provide heating, at least locally;
for example the CO peak labelled 1 coincides with the far-infrared source,
and the extension labelled 2 may be related to the star HD 147889. The
heating process is indirect: the grains absorb radiation, warm up, and
transfer energy to H_2 by collisions; CO is collisionally excited by H_2. If
the gas and grain temperatures have reached equilibrium, then the far-infrared emitting grains are quite cool, about 30 K. The embedded sources
also probably provide both the infrared radiation that is thought to pump
OH to produce the observed anomalous strength of the 1720 MHz and weakness
of the 1612 MHz satellite lines, and the ionizing radiation necessary to
create the C II regions in which the carbon radio-recombination lines
arise.

Given the high densities implied in this cloud, significant depletion
of heavy elements and the consequent growth of interstellar grains can be
expected within the finite lifetime of the collapsing cloud (§8.1). In the
hottest central condensation where $T_d \approx 30$ K, evaporation of a weakly bound
mantle might limit or reverse this process, but throughout most of the
cloud the grains should be sufficiently cool (see Fig. 8.1). Empirical
evidence consistent with grain growth has been obtained from a study of the
colours and wavelength dependence of polarization of several stars within
the extended dark cloud ($2°$) around HD 147889. Both the colour-excess
ratio E_{V-K}/E_{B-V} ($\approx 0.9R$) and λ_{max} increase with E_{V-K} (optical depth), and
the slope of the ultraviolet extinction curve for the brighter stars that
can be measured decreases with R. Thus larger grains are associated with
larger optical depths. However, the observations described above also suggested that the more optically thick regions are more dense, leading to the
conclusion that larger grains exist in more dense regions. The maximum
values of R and λ_{max} observed (4.5 and 0.8 μ respectively) for HD 147889
and its more obscured neighbour SR-3 imply that the grain size has grown by
a factor of 1.5. The corresponding mass increase, a factor of 3.4, may
have been limited by the availability of condensable material; the grains
would have begun with 30 per cent of the maximum, which is not unreasonable. Further support for this view has been provided by a subsequent
observation of a faint star coincident with one of the compact H II
regions. The spectrum and the radio flux indicate a B3-B5 main sequence
star, and hence $E_{B-V} \approx 2.9$ and $A_V \approx 12$ were derived. Though the obscuration is
much larger, the grain growth has not proceeded further ($R \approx 4.1$).

There is some contradictory evidence, however. When the variable
extinction method is applied to the OB stars in this region, including HD
147889 for which λ_{max} is largest, a well-determined, but normal, value of R

is obtained (3.1). This discrepancy has not been resolved.

Another hypothesis put forward to explain the evidence for grain growth is that the grain size distribution is altered by grain-grain collisions. The mean size increases as a result of coagulation of small grains, but the size is limited by the greater probability of large grains shattering in a collision (§11.3). Abnormally large values of λ_{max} have been discovered for highly obscured stars associated with several other regions of recent star formation, showing that the inferred grain growth is a common phenomenon.

Radiative transfer. The large amount of obscuration in dense interstellar dust clouds gives rise to theoretical problems in radiative transfer. Two of the most interesting are the penetration of ultraviolet radiation capable of dissociating molecules, and the radiation balance which determines the dust temperature in a shielded environment. Both of these have been treated in spherical geometry, as a first approximation, with either a Monte Carlo scattering method or numerical solution of the differential equations of radiative transfer.

The amount of ultraviolet penetration is strongly dependent on the adopted albedo. If the albedo is not zero, then the diffuse radiation field can be substantial. The penetration is further enhanced when the phase function is forward throwing (g>0). Theoretical models of dust scattering show a significant albedo even for absorbing materials, unless the grains are very small compared with the wavelength (§4.3). Furthermore there is direct evidence, from far-ultraviolet observations of scattered radiation, for an unexpectedly high albedo. Thus molecule lifetimes based on early calculations with zero albedo were severely overestimated, by factors as large as 10^4 in dense clouds ($A_V \approx 8$). Although the main uncertainties in these models are still $\tilde{\omega}$ and g, the complications of irregular density distributions and embedded sources should not be overlooked.

Determination of the temperature distribution of dust as a function of optical depth is subject to these same uncertainties. In addition the problem is strongly coupled at large optical depths, where re-radiated thermal radiation is the principal mode of energy transport. One series of models is illustrated in Fig. 9.10. The temperature decline with optical depth is not dramatic, since as we have seen a large change in the radiation field is required to produce even a small change in temperature. Nevertheless the temperature structure is important, because the values lie in the critical range for various surface phenomena (Fig. 8.1).

The gas in a dense molecular cloud cools principally through CO line radiation. If the column density is high and the velocity dispersion (or gradient) is low, the opacity in the lines is high and so the cooling radi-

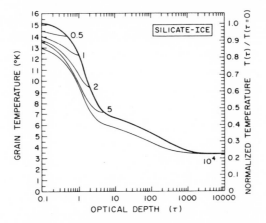

FIG. 9.10. Temperature distribution for dust within spherical
dark clouds. The free-space temperature for the ice-coated
silicate grains is about 15 K. The thin lines indicate the
reduction in temperature for clouds having the central optical
depths labelled. The thick line is the locus of central temperatures. (From C.M. Leung, Radiation transport in dense
interstellar clouds. I. Grain temperature, Astrophys. J.
199, 340 (1975). University of Chicago Press. ⊙1975 by the
American Astronomical Society. All rights reserved.)

ation is trapped. Then energy loss through collisions with the cooler dust
particles, a process which is always present, becomes relatively important
(eqn (6.7)). The grains can radiate away this extra energy with only a
slight change in temperature. When the grains are hotter than the surrounding gas, as a result of localized heating by an embedded stellar
source, the opposite process can take place, with collisions heating the
gas to the dust temperature. As mentioned above, this effect is suspected
in the ρ Oph dark cloud.

9.4. DUST ASSOCIATED WITH H II REGIONS

Dust is usually well mixed with neutral interstellar gas. As an interstellar cloud complex evolves towards star formation the dust-to-gas ratio
could rise above normal if the grains grow somewhat by accretion. At a
later stage, when one or more O stars form, part of the cloud containing
dust becomes an H II region. Grains within the ionized zone might be
altered in two ways, by sputtering in the hot gas, especially in the ionization front, or by evaporation when the grain temperature is raised by the
enhanced radiation field. The efficiencies of both of these destruction
mechanisms are uncertain as they depend on the (unknown) nature of the

grain material; conceivably the process would be selective, with loosely
bound icy grains or mantles being eroded away, leaving behind refractory
particles or cores of core-mantle particles. Sputtering is probably hin-
dered by the high grain charge established by the photoelectric effect near
the star; however, evaporation would be more important in the inner region
where the grains are hotter. Destruction of grains or their mantles would
decrease the dust-to-gas ratio to a normal or subnormal value. Outside of
the H II region the dust would be unaffected.

Internal dust surviving in the H II region alters the physics substantially and also produces several interesting observable phenomena.
Consider, for example, the effects on spectral lines. Differential extinction at optical wavelengths, by internal or external dust, affects observed
line ratios such as the Balmer decrement, thus complicating their interpretation. Particularly important are the effects on resonantly scattered
photons. Because the effective path length travelled by these photons can
be so much larger than a single traverse of the nebula, even a small optical depth of dust could provide efficient extinction. The infrared line of
He I at 10830 Å is affected in this way. The most important though is
Lyman α, since so much energy, one photon for virtually every Lyman continuum photon, ends up in this line. Even if Lyman-α photons are not absorbed
on their first passage through the nebula they are effectively backscat-
tered by surrounding H I (resonantly trapped), so that ultimately they are
destroyed by dust particles in the H II region. Absorption by dust also
substantially alters the excitation equilibrium, and consequently diagnostic indicators such as the Balmer decrement will be different.

Thermal re-radiation of the Lyman-α energy absorbed by the dust produces a powerful infrared source, whose characteristics contrast somewhat
with a thermal source heated directly by the stellar continuum. Because
grain heating is by a diffuse flux, the usual radial dilution is absent and
there is little temperature gradient throughout the ionized region. Thus
infrared emission would be seen from the entire H II region, without the
usual strong central peaking. This lack of temperature gradient naturally
affects the spectrum also. Furthermore, if there are too many grains, the
diffuse Lyman-α flux does not build up, and the grain temperatures are
depressed. Hence the curious situation arises in which the amount of
short-wavelength radiation ($\lambda < 10$ μ) can be inversely proportional to the
number of grains.

Given an amount of dust sufficient to absorb the stellar continuum,
an even more powerful infrared source can arise. If dust competes effectively for the Lyman continuum photons, however, the overall ionization
rate is decreased; the resulting smaller H II region is ionization-bounded
by dust, rather than gas. Furthermore, the size of the He ionization zone

relative to that of H is sensitive to any differential extinction (particularly absorption) in the far ultraviolet (E(504-912)). Thus the derivation
of the He/H abundance ratio from radio recombination lines becomes more
complicated.

There are potentially interesting dynamical effects of radiation
pressure on dust in the nebula. One is the induced drift of grains through
the gas which might separate smaller from larger grains; this is limited by
the strong electrostatic coupling of the particles to the gas. On the
other hand the contribution to the expansion of an H II region should be
important; we have already mentioned the observational evidence provided by
the Rosette Nebula and W3(A) (§8.3). Photoemission of electrons will be
preferentially on the side of the grain that faces the central star when
the grain size exceeds the wavelength of the photons absorbed, producing a
rocket effect which enhances the radial force on dust close to the star.
Evaluation of these effects depends on a detailed knowledge of the grain
properties, and should be reconsidered when the latter become more certain.

Extinction. With this brief introduction we can look at observational evidence for dust associated with H II regions. The presence of
dust is inferred from the lack of correspondence between Hα and radio
brightness distributions. Some H II regions are in fact completely
obscured optically by surrounding dust, and are detectable only by their
radio and infrared emission. Dust can be seen directly in dark obscuring
lanes, elephant trunks, and globules. Dense globules actually within or at
the periphery of the ionized region are distinguished by a bright rim (in
Hα). Embedded globules not seen in silhouette are revealed by photographs
in the light of low excitation emission lines (like [N II]) which are
enhanced where ionization fronts penetrate the dense neutral regions. Line
splitting and other velocity features have been attributed to this same
on-going ionization phenomenon.

While some of the extinction found by any technique can be foreground, it can often be argued, for instance by the strong variability
within an H II region, that the dust is closely related to the source. A
number of methods are used to derive the amount of extinction. A partial
extinction curve can be derived from the observed Balmer decrement if the
theoretical prediction is reliable, or in the usual fashion (§3.1.1) an
extinction curve can be derived for the illuminating star(s). There are
two complications. Scattered light may be included along with stellar
light if the observing aperture is too large, so that the contribution of
scattering to extinction is diminished. Such an effect is a complex function of the geometry, optical depth, phase function, and albedo, and has
been modelled only for some idealized cases. Second, the star may have an

infrared excess from circumstellar material, which will artificially
steepen the extinction curve at long wavelengths. This problem has been
encountered in the Orion Nebula. Photometric and polarimetric observations
now indicate that the extinction law is normal, except for a few stars for
which both R and λ_{max} are large.

The absolute amount of visual extinction can be measured independently by comparing observed Balmer line strengths (usually Hα or Hβ) with
the value expected on the basis of radio brightness; thus the ratio of
total to selective extinction can be estimated. For example, this method
applied to the extended radio component W3(A) of the W3(continuum) complex
of H II regions gives $A_V > 14$. Associated with this component is an extended
infrared source (W3-IRS 1); maps at 2.2 and 20 μ show it has an angular
extent comparable with the radio source (∼40 arc sec) and a similar shell
structure. The spectrum of this source is shown in Fig. 9.11. The broken
curve shows the predicted infrared radiation from free-free, bound-free,
and bound-bound transitions in the H-He plasma. Two important points can
be emphasized. Near 10 μ the infrared radiation detected exceeds the gaseous contribution by a large amount; thermal emission by hot dust is responsible. In the near infrared (λ<2 μ), however, the observations fall below
the expectation, indicating that some extinction is present. The amount of
extinction can be found in two ways: directly, by comparing the observed

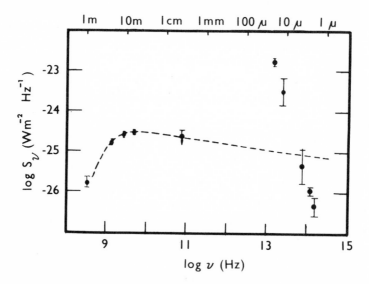

FIG. 9.11. Energy distribution of the source W3(A)/IRS 1. The
broken curve extrapolates the thermal emission of the gas
towards higher frequencies. (After C.G. Wynn-Williams, E.E.
Becklin, and G. Neugebauer, Infra-red sources in the H II
region W3, Mon. Not. R. astron. Soc. 160, 1 (1972).)

and predicted fluxes, or indirectly, by deriving a near-infrared colour
excess and assuming the shape of the extinction curve (a near-infrared
ratio of total to selective extinction). W3-IRS 1 has $\tau(2.2\ \mu)\approx1.3$, which
yields $A_V\approx14$ for normal-sized grains (this last step is a weak link).
Finally spectral absorption features, detected at 3.1 and 10 μ against the
strong continuum of some compact thermal dust sources, provide another measure of A_V. Adopting a silicate band strength of 5000 cm^{-1} and normal-
sized grains leads theoretically to $A_V\approx12\tau(9.7\ \mu)$, a ratio similar to that
found empirically for some sources (§10.2). Since $\tau(9.7\ \mu)\approx2.2$ for W3-IRS
1, $A_V\approx26$; this higher value may simply be an overestimate or may refer to a
somewhat more embedded hot central source. In any case many compact components of H II regions are undoubtedly very highly obscured. The required
large amount of material had its origin in the cool molecular cloud in
which the hot stars are forming. At this stage of evolution the distribution is highly patchy owing to fragmentation; for example, W3-IRS 5, a more
compact source only 30 arc sec away from W3-IRS 1, has $\tau(9.7\ \mu)\approx7.6$!

Scattered light. Diffuse continuous optical radiation has been measured in many H II regions at a level much higher than can be explained by
bound-free and two-photon emission from the gas. Light scattering by dust
in the H II region is responsible. Direct confirmation of a scattering
origin is provided by the detection of the 4686-Å photospheric absorption
line of He in the reflected light in the Orion Nebula and by the discovery
of linear polarization (like a reflection nebula) in several H II regions,
including Orion. More extensive polarization (and colour) studies should
be possible; their interpretation would be more straightforward than for
most reflection nebulae provided that there is a more nearly spherically
symmetric geometry.

The spatial distribution of reflected light need not follow that for
the emission lines (or an atomic continuum) even if the dust and gas are
well mixed. In the first place, emission by the gas depends on the square
of the density and so is much more affected by density gradients and inhomogeneities. Furthermore the scattered light is more centrally peaked
because of the r^{-2} dependence of the illumination. As an illustration consider the Rosette Nebula, which is well known for its ring-shaped appearance in Hα; in contrast with this, diffuse light is observed filling in the
central portion (cf. Barnard's Loop in §5.3). The Hubble relation for the
extent of a reflection nebula (eqn (5.1)), combined with the usual results
for the size of a Stromgren sphere, shows that scattered light will generally be confined to the inner part of an H II region, unless the gas density is very high.

The different behaviour of the surface brightness of Hβ with respect

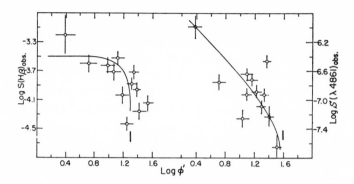

FIG. 9.12. Hβ and 4861-Å continuum observations of the surface brightness of M 8. The thin lines represent a uniform density model. Limiting size of the scattering cloud is somewhat larger than the ionized region, as indicated by the vertical bars. (From C.R. O'Dell, W.B. Hubbard, and M. Peimbert, Photoelectric photometry of gaseous nebulae. III. Scattered light in three bright H II regions, Astrophys. J. 143, 743 (1966). University of Chicago Press. ⁰1966 by the University of Chicago.)

to that of the optical continuum is illustrated in Fig. 9.12 for the Lagoon Nebula (= NGC 6523 = M 8). The solid curves show the predictions from a simple uniform sphere model. From the relative scales a measure of the gas-to-dust ratio is determined: $N_H/A_\lambda\tilde{\omega} \simeq 2 \times 10^{21}$ cm^{-2} mag^{-1} at Hβ. A more detailed model allowing for a density gradient in the source gives similar results at all radii. Apparently the proportion of dust is not very different from the average for neutral regions of the interstellar medium (eqn (9.2)). However, a similar analysis of the Orion Nebula suggests an under-abundance of dust (1.4×10^{22} cm^{-2} mag^{-1}) in the inner region ($\sim$1 arc min) and possibly an over-abundance (5×10^{20} cm^{-2} mag^{-1}) in the outer region. It is possible that part of the apparent large change in abundance may be accounted for by a change in albedo. For example, grains retaining their icy mantles in the outer region would have a relatively high albedo, whereas in the inner region the mantle could be evaporated leaving a smaller refractory particle with a lower albedo. Another serious difficulty with any interpretation is the sensitivity of the result to inhomogeneities in the density distribution. Because of the different density dependences of the brightness mentioned above, the gas-to-dust ratio would be overestimated by a factor $f^{-\frac{1}{2}}$, where f (<1) is the filling factor for a clumpy distribution. Values of f as low as 0.03 have been suggested by emission-line observations of some nebulae, including the inner region of the Orion Nebula.

Thermal emission. Infrared observations provide an important means
of investigating the dust in and around H II regions. One interesting discovery is that the luminosity measured in the 40-350 μ range, which approx-
imates the total infrared luminosity, is well correlated with the ionization rate in the nebula as calculated from radio flux density measurements
at centimetre wavelengths (Fig. 9.13). This correlation suggests that
non-ionizing sources of heating are unimportant. To clarify the relationship of the infrared luminosity to the energy available from the central
stars two loci have been plotted in the figure, one for the zero-age main
sequence bolometric luminosity and one for the portion which is converted
into Lyman-α radiation in an optically thin H II region. Thermalization of
trapped Lyman-α radiation, which would require only a relatively low optical depth, is clearly not sufficient; in fact, less than one-third of the
energy can have been converted in this fashion. The rest of the energy can
be absorbed in two distinct ways. If the optical depth in the H II region
is low, then the stellar continuum on the long-wavelength side of the Lyman
limit (about one-third of the energy) and the diffuse line and continuous
radiation produced by degradation of the Lyman continuum in the H II region
(another third) must be absorbed by dust in the surrounding H I region.
Thus the infrared source would be more extended than indicated by the radio

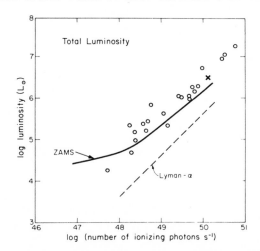

FIG. 9.13. Total luminosity of H II regions, from 40 to 350 μ,
as a function of the ionization rate. Main sequence luminosity
from spectral type B0 to 04 is shown by the ZAMS line. The
fraction of energy at Lyman α is also indicated. The cross is
for M 17. (After C.G. Wynn-Williams and E.E. Becklin, Infrared
emission from H II regions, Publ. astron. Soc. Pac. 86, 5
(1974). Courtesy of the Publications of the Astronomical
Society of the Pacific.)

continuum maps of the H II region. Furthermore, these grains would be
cooler than the component within the H II region, giving rise to a broadening of the integrated infrared spectrum and a wavelength dependence of the
size of the source. If there were depletion of grains within the H II
region, a shell source would be seen at longer wavelengths.

Alternatively the energy could be absorbed within the H II region, if
there were sufficient dust. Because internal dust would absorb the ionizing continuum, the size of the H II region would be limited and would correspond to the volume emitting the bulk of the infrared radiation. Similarly, the observed He II/H II ratio could be depressed by selective
extinction in the far-ultraviolet. Since the overall ionization rate would
be reduced, points in Fig. 9.13 would lie to the left of the main sequence
locus, giving the appearance of over-luminous sources for a given ionization rate. As a result of the temperature gradient which accompanies
direct absorption of starlight, near-infrared emission would be quite
strongly centrally peaked.

M 17. As an illustration of the kinds of data available for
interpretation we shall consider briefly the bright H II region M 17 (= NGC
6618 = the Omega Nebula = the Horseshoe Nebula = the Swan Nebula; see Fig.
9.5). Only a few other nebulae, like Orion and its associated molecular
cloud OMC 1, have been studied as extensively. Optically M 17 is highly
obscured; the peak of line emission is considerably offset from the two
main peaks of radio continuum emission (Figs. 9.14 and 9.16). Comparison
of the isophotes reveals $A_V \approx 4$ in the optically bright region and $A_V \approx 7$ near
the radio peaks. One O star observed in the nebula has $A_V > 8$, and the other
exciting stars are not seen optically. Although scattered light has been
detected as well, it has not been mapped.

The derived properties of the brighter (fainter) radio component seen
in Fig. 9.14 are density 950 (640) cm^{-3}, diameter 2.7 (3.5) pc, and mass
220 (330) M_Θ. Together the radio peak sources account for only half of the
integrated brightness. The less dense extended component therefore com-
prises much more mass.

The cross marked for M 17 in Fig. 9.13 shows that in this source, as
in the other sources, there is complete conversion of the stellar luminosity into infrared radiation. The unusual breadth of the infrared spectrum
suggests the existence of a wide range of grain temperatures. This would
arise naturally if the near-infrared radiation arose from hot grains within
the H II region while longer-wavelength radiation was emitted by cooler
grains further out in the H I gas. However, infrared maps do not support
this picture. The distributions at 10, 20, and 100 μ (Figs. 9.15 and 9.16
for the latter two wavelengths) are all very similar to the radio maps

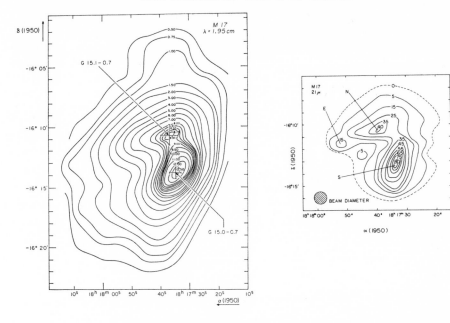

FIG. 9.14. Intermediate resolution (2 arc min) map of M 17 at 1.95 cm, showing two main peaks of radio emission. The contour units are antenna temperature in K (about half the brightness temperature). (From J. Schraml and P.G. Metzger, Galactic H II regions. IV. 1.95 cm observations with high angular resolution and high positional accuracy, Astrophys. J. 156, 269 (1969). University of Chicago Press. ©1969 by the University of Chicago. All rights reserved.)

FIG. 9.15. A 21-μ map of the central region of M 17 (aligned with Fig. 9.14 in declination and to the same scale). A third source E is resolved. The contour units are about 10^{-17} W m^{-2} Hz^{-1} $sterad^{-1}$. (From D. Lemke and F.J. Low, 21-micron observations of H II regions, Astrophys. J. Lett. 177, L53 (1972). University of Chicago Press. ©1972 by the American Astronomical Society. All rights reserved.)

(when allowance is made for the differing beam diameters). This is closer to the dust-limited H II region concept, and is consistent with the displacement of the data point in Fig. 9.13. However, towards the south-west, where there is a rapid decline in the radio continuum, there is an extension of the 100-μ emission which may be related to the neutral gas detected there by the peak in CO molecular line emission (Fig. 9.17).

Molecular emission from CS and H_2CO also peaks in the same position. The derived molecular hydrogen density is 3×10^4 cm^{-3} and the mass is

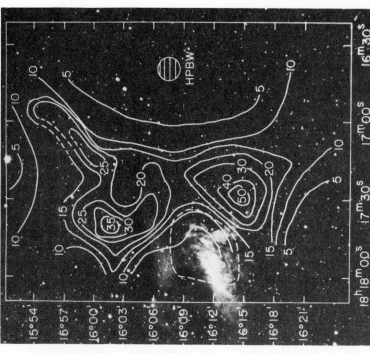

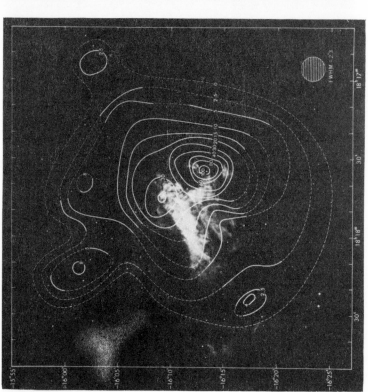

FIG. 9.16. A 100-μ map of M 17 showing a peak coinciding with the main source at 20 μ and 1.95 cm but extending somewhat more to the south-west. The offset of the optical emission is due to obscuration. (From D.A. Harper, F.J. Low, G.H. Rieke, and H.A. Thronson, The infrared emission of M 17, Astrophys. J. 205, 136 (1976). University of Chicago Press. ©1976 by the American Astronomical Society. All rights reserved.)

FIG. 9.17. M 17 map of the apparent brightness temperature of ^{12}CO at 20 km s^{-1} (broken contour at 23 km s^{-1}). (From C.J. Lada, Detailed observations of the M 17 molecular cloud complex, Astrophys. J. Suppl. 32, 603 (1976). University of Chicago Press. ©1976 by the American Astronomical Society. All rights reserved.)

about 10^4 $M_\odot$. Coincident with this dense neutral cloud are two groups of variable H_2O molecular masers, and an infrared (10 and 20 μ) point source. The total mass of the CO cloud is estimated to exceed 3×10^4 $M_\odot$, including the bright spot to the north (2×10^3 $M_\odot$) where excess 100-μ emission also occurs. Furthermore, the CO cloud shown in Fig. 9.17 is itself only the north-east end of a flattened cloud, which extends more than 1° (85 pc) along the Galactic plane (see Fig. 9.5) and has a mass of about 10^6 $M_\odot$. These observations help place the bright H II region in proper perspective; it appears to be a relatively small fragment of a much larger molecular cloud.

To explain a range in grain temperature in excess of that produced by the usual radial gradient the existence of more than one species of grain, with different far-infrared emissivities (and hence temperatures), has been postulated. The difference in composition might be related to the presence or absence of an icy grain mantle on a refractory core; sublimation of the grain mantle would certainly occur for grains hot enough to emit efficiently at 10 μ. More sublimation would be expected near the stellar sources, thus helping to reduce the central peaking in the short-wavelength maps, which is not much more pronounced than in the 100-μ maps. The amount of central peaking is probably also strongly influenced by density gradients, not considered in simple models.

Alternatively, it can be supposed that there are neutral globules embedded throughout the H II region. Grains within the globules, being shielded from the Lyman continuum radiation, would be cooler and yet could exist in relative abundance close to the ionizing stars, explaining the similar brightness distributions at 10 and 100 μ. (Considerable Lyman-α heating could occur throughout the less dense regions and in the ionization fronts.) At the same time the spectrum would be broadened, being determined largely by the temperature distribution within a single globule.

Planetary nebulae. A discussion of dust in planetary nebulae in many ways parallels that for dust in H II regions, and so only brief mention will be made here. A common property among planetary nebulae is powerful thermal infrared emission, which reveals the presence of large amounts of dust. In some nebulae optical continuum radiation in excess of a gaseous contribution has been attributed to scattering by dust. Maps of polarization would be useful. The effects of extinction by dust on optical and ultraviolet line strengths have been examined as well.

The infrared luminosity indicates a substantial conversion of ultraviolet energy. Lyman-α heating is energetically sufficient in some objects, but in many others direct heating by the stellar continuum must be occurring. Direct heating would also account for the nebulae in which the

10-μ brightness distribution is more centrally concentrated than the optical and radio emission. Other infrared maps show the same ring structure
(central hole) present in radio and optical data, but the question of dust
outside the ionized region is not settled.

As in H II regions, derivation of the total mass of dust from
infrared spectra is complicated by the broad range of grain temperatures,
which arises both as a radial gradient and, at any radius, from various
grain sizes and compositions. Thin-shell single-temperature models are
unrealistic. Furthermore, there are uncertainties in the grain composition
and size distribution and in the complex refractive indices of grain materials. Order-of-magnitude estimates give about $10^{-3} M_{\odot}$ of dust and a
gas-to-dust mass ratio less than 10^3.

Infrared spectral features are discussed in §10.2. In H II regions
prominent emission or absorption by silicates can be seen at 10 μ. This
feature and the 3.1-μ ice band detected in sources associated with dense
molecular clouds are charactersitics of the composition of interstellar
grains. Grains found in planetary nebulae are quite distinct. The 10-μ
feature is never seen, which is suggestive of a carbon-rich environment
since silicates are condensates in oxygen-rich envelopes. Graphite is the
main condensate in the carbon-rich case (§11.1.2). Unfortunately graphite
produces a featureless continuum. Indirect support for graphite comes from
a crude mass estimate based on the substantial infrared excess, which
requires that grains be composed of more than the trace elements. Although
the materials responsible for the unidentified emission features at 3.3,
3.4, 6.2, 7.7, 8.7, and 11.3 μ may account for only a small fraction of the
infrared emission and the total grain mass, their identification will be
important to understanding the chemistry of the condensation process.

Some models of the evolution of planetary nebula envelopes suggest
that radiation pressure on dust particles, which drag the gas with them, is
responsible for creating the central hole and maintaining its relatively
sharp inner boundary. Grains which can form at an early stage are most
important; graphite could begin condensing at about 1700 K. If graphite
were the dominant grain material, then the immediate precursors of planetary nebulae would be late-type carbon stars. In fact some IRC and CRL
sources have been proposed as various stages of pre-planetary nebulae
(§7.2.1).

We note that there are qualitative similarities in the formation of
dust in novae and planetary nebulae.

9.5. INTERSTELLAR DUST IN THE SOLAR SYSTEM

A growing body of evidence on the nature of interplanetary dust particles

in the zodiacal cloud indicates that any direct contribution by penetration
of interstellar dust into the solar system is very small. On the other
hand there may be an important indirect link, since interplanetary particles are believed to originate as cometary debris, and cometary dust particles themselves may have grown by agglomeration of much smaller grains of
interstellar dust in the early stages of the solar system. Some features
of interplanetary dust will therefore be reviewed.

Interplanetary dust. The optical depth of the zodiacal cloud is so
low as to make extinction negligible. However, the illumination by the sun
is sufficiently intense to produce detectable scattered light, called the
zodiacal light; until recently this provided the only means for studying
interplanetary particles. Zodiacal light derives its name from its concentration in the mean plane of the solar system (near the ecliptic plane).
As in reflection nebulae, its brightness and polarization distributions are
determined by the spatial distribution of particles and their phase functions. Measurements as a function of elongation indicate a density dependence on heliocentric distance which falls approximately as $R^{-1.5}$ and cuts
off at 3.3 AU. The axial ratio of the highly flattened cloud is about 7.
Two distinct features of the zodiacal light are the Gegenschein, produced
by backscatter from particles in the anti-solar direction, and the inner
zodiacal light or F corona, which arises both from particles near the sun
and from small-angle diffraction by particles much closer to the observer.

Extensive ground-based observations of colour and polarization, supplemented by measurements from spacecraft, suggest that the scattering particles are much larger than interstellar grains, sizes of 10-30 μ (10^{-9} g)
being typical. Particles seem to be roughened or irregular in shape. An
increased scattering optical depth in the far-ultraviolet might indicate
the presence of a large number of submicron-sized grains.

Thermal infrared emission from warm interplanetary dust has been
detected in the ecliptic. In the F corona, where the grains are hotter,
the emissivity is even higher. Enhanced thermal radiation also comes from
comets that release copious amounts of dust. A 10-μ spectral feature
detected in several comets is similar to that produced by circumstellar
dust and possibly has the same origin (§10.2.2). This peak has also been
identified tentatively in the F corona.

Meteoritic phenomena are associated with much larger interstellar
particles. Only rare massive ($>10^3$ g) bodies survive passage through the
earth's atmosphere. Of these the primitive carbonaceous chondrites have
compositions that are thought to be closest to that of zodiacal light particles, though there may not be a direct relationship. Common visual meteors result from less massive (1 g) but more numerous meteoroids (princi-

pally like chondrites); shower meteors are debris from comets. Even more
abundant are smaller (10^{-5} g) meteoroids, whose ionization trails in the
earth's atmosphere are detectable by radar; meteoroids of the smallest size
overlap with the largest zodiacal light particles (10^{-7} g). A break and
plateau in the power-law size distribution below 10^{-5} g (the mass of particles contributing most to the total of 3×10^{19} g in the zodiacal cloud)
suggest that the more massive meteoroids are more closely related to primary particles from comets, whereas the less massive grains are secondary,
resulting from various destructive processes.

In one attractive scheme particles of mass exceeding 10^{-6} g decay by
mutual collisions, producing smaller fragments. Smaller meteoroids, losing
orbital angular momentum according to the Poynting-Robertson effect, spiral
in towards the sun; at heliocentric distances of less than 1 AU the particles begin to be broken down by sputtering, sublimation, and mutual collisions. As the characteristic size decreases further, however, the ratio
of radiation pressure force to gravity (eqn (8.24)) rises, leading to an
increase in orbit size. If β exceeds unity (depending on the details of
the complex refractive index) the small grains are expelled rapidly. The
mass turnover time scale is estimated to be about 10^5 y, so that an injection of new material at a rate 10^7 g s^{-1} is required to maintain the cloud.
This can be supplied by comets. A balance between the input of fresh
material (size > 100 μ or 10^{-6} g) and subsequent breakdown and ejection
sets up size and radial distributions at least qualitatively similar to
those observed. More direct evidence is the detection, using spacecraft
sensors, of micrometeoroids streaming outwards from the inner solar system.
On the reasonable hypothesis that they represent the last stage in the evolutionary scheme outlined above, they have been called β-meteoroids.

The nature of micrometeoroids. A distinct advantage over interstellar studies is the direct detection of individual interplanetary particles
by various spacecraft (one or two interstellar particles may have been
intercepted), and such historical evidence as is provided by microcraters
in lunar surface material. However, of greatest potential importance is
the collection of micrometeoroids high in the stratosphere (Fig. 9.18),
which allows subsequent chemical analysis. About 100 particles with sizes
in the range 2-20 μ have been identified as extraterrestrial by the observation that the overall composition of Mg, Fe, Si, S, Ca, and Ni resembles
the pattern of solar abundances. The micrometeoroids are actually aggregates of submicron-sized grains which themselves appear to be made up of
many smaller particles (size $\simeq$ 50-500 Å). Inclusions of Ni-bearing iron
sulphides and iron-poor silicates (olivine and pyroxene) appear to have
characterized the larger bodies from which these aggregates derive. Such

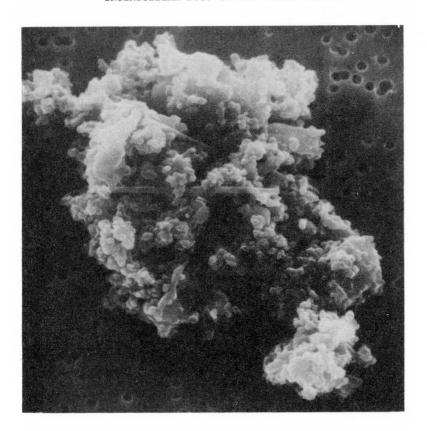

FIG. 9.18. Scanning electron micrograph of a micrometeoroid reveals a loose aggregate of irregular submicron-sized particles. Peculiar rod-like components are found in some other aggregates. Chemical composition is characteristic of type C1 carbonaceous chondrite material. This particular specimen is U2-11C(7), collected at 20 km altitude with a NASA U-2 aircraft. Scale: 1 μ = ———. (University of Washington photograph, courtesy of D.E. Brownlee.)

structure is reminiscent of the fine-grained material that forms type C1 and the matrix of type C2 carbonaceous chondrites (although on a 100-Å scale the micrometeoroids are structurally distinct), and suggests an origin under similar (though not common) physical conditions.

If it is supposed that the refractory components (cores) of interstellar grains survive in the outer regions of the primitive solar nebula, then the finest grains seen in cometary micrometeoroids and perhaps those in primitive chondrites might be identified with the remnants of interstellar grains. We note, however, that the amount of carbon present is much lower than expected if the interstellar extinction bump at 2200 Å is to be

explained by graphite (§10.2.1). Carbon occurs mostly in the form of high
molecular weight organic (tarry) compounds in carbonaceous chondrites
(there are traces of graphite) and as finely divided carbon in the micrometeoroids. To address this discrepancy it has been suggested that through
surface reactions in the solar nebula much of the graphite might have been
converted into volatile hydrocarbons and subsequently have been lost (some
of the products would be trapped within grain clumps). Note that the dominant form of carbonaceous chondrite material, hydrated silicates, might not
be exactly representative of the original interstellar grains either, since
hydration might have occurred in the solar nebula at the time of accumulation of the meteoroids. (Subsequent heating of meteoroids might produce
further metamorphosis.) Additional evidence bearing on the survival of
interstellar grains will be mentioned below.

Consider an alternative, that the submicron-sized components of the
micrometeoroids and chondrites originated in the solar nebula. Equilibrium
condensation, as the solar nebula cools from a high temperature at which
all material is vaporized, has been shown to provide an explanation for
chemical fractionation in the inner solar system (§11.1.2). The question
arises whether this hypothesis is relevant in the outer regions where comets and meteorites formed. Certainly there is evidence in the details of
the structure and mineralogy of ordinary chondrites and their inclusions
for chemical fractionation and the expected products of equilibrium condensation. On the other hand, there is little evidence in primitive (C1) carbonaceous chondrites for abundance anomalies associated with depletion in
high-temperature condensates. Rather it appears that the layer-lattice
silicates (and their magnetite inclusions) could have condensed directly
from the gas phase at relatively low temperatures ($\sim$350 K), and thus have
preserved the cosmic pattern of abundances; the same statement holds for
the micrometeoroids. The possibility of forming large quantities of submicron-sized particles of solar composition in the solar nebula is thereby
demonstrated. If in addition much of this material were ultimately dispersed into the interstellar medium, probably in finely divided form, solar
nebulae would represent a potentially important source of interstellar
grains.

Regardless of which of these fundamentally distinct hypotheses pre-
vails, it seems plausible that the primitive silicate composition of the
submicron-sized particles making up the observed aggregates is also a char-
acterisitic of refractory interstellar grain material.

Isotopic anomalies in meteorites. Some interstellar grains may have
formed in regions with anomalous (non-solar) isotopic composition. If
these are among the grains present in the cloud which becomes the solar

nebula, then they represent an initial inhomogeneity. Aggregation of the
individual grains into larger interplanetary meteoroids would have preserved the anomalies, though with varying degrees of dilution by material
of normal composition. The general failure to detect widespread isotope
anomalies in meteoritic material does not necessarily mean that interstellar grains were completely vaporized and homogenized in the gas phase,
because grains of the requisite anomalous composition might not even exist.

Of greatest interest here are the O and Mg anomalies observed in the
inclusions (themselves aggregates of relatively high temperature ($\sim$1500 K)
condensates) in carbonaceous chondrites. The O anomalies appear as small
deficiencies (compared with solar abundances) of ^{18}O and ^{17}O relative to
^{16}O. Each of the heavy isotopes is depleted by the same amount (which
establishes that the fractionation is not chemical), but from inclusion to
inclusion this amount varies, suggesting a mixing of an inhomogeneity containing virtually pure ^{16}O with different quantities of solar-composition
material. In this context it is interesting to note that the matrix material (lower-temperature condensates) of carbonaceous chondrites has an isotope composition lying at the normal-composition end of this mixing line.
Isotopic homogeneity within different phases of the meteoritic inclusions,
as well as the appearance of the crystals, indicates that the inclusions
have been partially melted and recrystallized, possibly into different mineral forms. Thus exact characterization of the interstellar particles
which carried the anomaly is not possible. We note that material enriched
in ^{16}O occurs in the zone of explosive C and O burning in type 2 supernovae, and could therefore have been present in the pre-solar cloud as interstellar dust formed from a cooling supernova remnant.

This scanty evidence for the survival of interstellar particles (anomalous and normal) in somewhat modified form must be confronted with the
results of radioactive dating, which place the time of accumulation of the
meteorites at 4.6×10^9 y, i.e. the origin of the solar system, not earlier. Possibly the ^{87}Rb and ^{235}U are loosely bound in the volatile mantles
of interstellar grains, which do vaporize and later recondense.

The enrichment of ^{26}Mg in the same type of inclusion is somewhat different. Examination of the $^{26}Mg/^{24}Mg$ ratio in several distinct mineral
phases of a coarse-grained inclusion revealed a tight correlation with
$^{27}Al/^{24}Mg$. The same type of Al-Mg isochron has been discovered within a
single crystal in which the $^{27}Al/^{24}Mg$ ratio changes with position. The
explanation seems to be as follows: condensation occurred in a gas
enriched with ^{26}Al; chemical fractionation during equilibrium condensation
at this time led to minerals generally enriched in both isotopes of Al;
subsequent β decay of ^{26}Al left ^{26}Mg in these minerals, in proportion to
the amount of stable ^{27}Al. Any subsequent melting and recrystallization

(with further fractionation), for which there is evidence, must have occurred before ^{26}Al decayed, or the isochron would not be observed. However,
the half-life of ^{26}Al is a mere 7.4×10^5 y; therefore condensation must
have followed soon after the nucleosynthetic event which produced ^{26}Al,
probably explosive carbon burning in a type 2 supernova. This requirement
has led to the hypothesis that passage of the remnant of a nearby supernova
actually triggered the collapse of the solar nebula, and that grains formed
in the supernova remnant carried the ^{26}Al anomaly into the solar system in
that short time interval. Injection in the gaseous phase is possible but
not favoured, because it would require condensation on a gas-mixing time
scale, which is very rapid. Enrichment of ^{16}O may have been produced by
grains formed during this same event, but because of the lack of similar
chronological restrictions an origin in interstellar grains from earlier
supernova events cannot be ruled out. If the supernova hypothesis is correct, then it should be possible to detect isotope anomalies in several
other common elements.

The evidence from carbonaceous chondrites, generally considered to be
the least processed of the meteorites, does not seem to favour survival of
the interstellar grains in their original form; in fact much of the refractory material appears to have been vaporized completely. On the other
hand, if isotope anomalies are carried by solid particles then at least
some grains must not have been vaporized; melting, with accompanying metamorphosis but not equilibration with the gas, is common in this case. Cometary micrometeoroids may therefore present the only hope of examining
intact interstellar grain cores, if comets are in fact unprocessed accumu-
lations of interstellar dust and molecules.

Independent of the resolution of these detailed problems, it appears
that many submicron-sized particles have been able to form in the prevail-
ing conditions in the solar nebula. The close correspondence of their mineralogy to the predictions of equilibrium condensation theory indicates
that this approach can be a useful guide to condensation in other nebulae
(§11.1.2).

FURTHER READING

§9.1

HOLMBERG, E.B. (1974). Distribution of clusters of galaxies as related to
        galactic absorption. Astron. Astrophys. 35, 121.
HEILES, C. (1976). An almost complete survey of 21-cm line radiation for
        |b|⩾10°. III The interdependence of H I, galaxy counts, reddening,
        and galactic latitude. Astrophys. J. 204, 379.
TEERIKORPI, P. (1978). A study of Galactic absorption as revealed by the

Rubin et al. sample of Sc galaxies. Astron. Astrophys. 64, 379.

LUCKE, P.B. (1978). The distribution of colour excesses and interstellar
reddening material in the solar neighbourhood. Astron. Astrophys.
64, 367.

§9.2

See §9.1. (Heiles 1976).

HEILES, E.B., and JENKINS, E.G. (1976). An almost complete survey of 21-cm
line radiation for |b|⩾10°. V Photographic presentation and qualita-
tive comparison with other data. Astron. Astrophys. 46, 333.

BOHLIN, R.C. (1975). Copernicus observations of interstellar absorption at
Lyman alpha. Astrophys. J. 200, 402.

SERKOWSKI, K. (1968). Correlation between the regional variations in the
wavelength dependence of interstellar extinction and polarization.
Astrophys. J. 154, 115.

HEILES, C. (1974). A modern look at 'interstellar clouds'. In Galactic
radio astronomy (eds. F.J. Kerr and S.C. Simonson). Reidel, Dordrecht.

NEE, S.F., and JOKIPII, J.R. (1978). Interstellar polarization in an
irregularly fluctuating medium. I. Basic theory. Preprint.

MARTIN, P.G., and CAMPBELL, B. (1976). Circular polarization observations
of the interstellar magnetic field. Astrophys. J. 208, 727.

§9.3

LYNDS, B.T. (1968). Dark nebulae. In Nebulae and interstellar matter
(eds. B.M. Middlehurst and L.H. Aller). University of Chicago Press,
Chicago.

GOUDIS, C., and JOHNSON, P.G. (1978). The obscuration and excitation of
the North American (NGC 7000) and Pelican (IC 5070) nebular complex.
Astron. Astrophys. 63, 259.

FITZGERALD, M.P., STEPHENS, T.C., and WITT, A.N. (1976). Surface brightness profiles of dark nebulae: the Thumb Print Nebula in Chamaeleon.
Astrophys. J. 208, 709.

TUCKER, K.D., DICKMAN, R.L., ENCRENAZ, P.J., and KUTNER, M.L. (1976). The
relation between carbon monoxide emission and visual extinction in
cloud L134. Astrophys. J. 210, 679.

DICKMAN, R.L. (1977). Bok globules. Sci. Am. 236 (6), 66.

BOK, B.J., and McCARTHY, C.C. (1974). Optical data for selected Barnard
objects. Astron. J. 79, 42.

BOK, B.J. (1977). Dark nebulae, globules and protostars. Publ. astron.
Soc. Pac. 89, 597.

ZUCKERMAN, B., and PALMER, P. (1974). Radio radiation from interstellar molecules. <u>A</u>. <u>Rev</u>. <u>Astron</u>. <u>Astrophys</u>. 12, 279.

RICKARD, L.J., PALMER, P., BUHL, D., and ZUCKERMAN, B. (1977). Observations of formaldehyde absorption in the region of NGC 2264 and other Bok globules. <u>Astrophys</u>. <u>J</u>. 213, 654.

BROWN, R.L., and ZUCKERMAN, B. (1975). Compact H II regions in the Ophiuchus and R Coronae Austrinae dark clouds. <u>Astrophys</u>. <u>J</u>. <u>Lett</u>. 202, L125.

CARRASCO, L., STROM, S.E., and STROM, K.M. (1973). Interstellar dust in the rho Ophiuchi dark cloud. <u>Astrophys</u>. <u>J</u>. 182, 95.

BERNES, C., and SANDQVIST, A. (1977). Anisotropic scattering in dark clouds and formaldehyde lifetimes. <u>Astrophys</u>. <u>J</u>. 217, 71.

SPENCER, R.G., and LEUNG, C.M. (1978). Infrared radiation from dark globules. <u>Astrophys</u>. <u>J</u>. 222, 140.

§9.4

MUNCH, G., and PERSSON, E. (1971). The distribution of gas and dust in the Orion Nebula. <u>Astrophys</u>. <u>J</u>. 165, 241.

SARAZIN, C.L. (1977). Effects of dust on the structure of H II regions. <u>Astrophys</u>. <u>J</u>. 211, 772.

ZEILIK, M. (1977). Two-component dust models of near-infrared emission from compact H II regions. <u>Astrophys</u>. <u>J</u>. 213, 58.

AANNESTAD, P.A. (1978). Models of infrared emission from dusty and diffuse H II regions. <u>Astrophys</u>. <u>J</u>. 220, 538.

ELLIOT, K.H., and MEABURN, J. (1974). Observations of major ionization fronts in M 42 and M 8. <u>Astrophys</u>. <u>Space</u> <u>Sci</u>. 28, 351.

BREGER, M. (1977). Intracluster dust, circumstellar shells, and the wavelength dependence of polarization in Orion. <u>Astrophys</u>. <u>J</u>. 215, 119.

HARRIS, S., and WYNN-WILLIAMS, C.G. (1976). Fine radio structure in W3. <u>Mon</u>. <u>Not</u>. <u>R</u>. <u>astron</u>. <u>Soc</u>. 174, 649.

WILLNER, S.P. (1977). 8 to 13 micrometer spectrophotometry of compact sources in W3. <u>Astrophys</u>. <u>J</u>. 214, 706.

HACKWELL, J.A., GEHRZ, R.D., SMITH, J.R., and BRIOTTA, D.A. (1978). Infrared maps of W3 from 4.9 to 20 microns. <u>Astrophys</u>. <u>J</u>. 221, 797.

PALLISTER, W.S., PERKINS, H.G., SCAROTT, S.M., BINGHAM, R.G., and PILKINGTON, J.D.H. (1977). Optical polarization in the Orion Nebula. <u>Mon</u>. <u>Not</u>. <u>R</u>. <u>astron</u>. <u>Soc</u>. 178, 93P.

MATHIS, J.S. (1971). Internal dust in gaseous nebulae. II. Absorption of Lyman-continuum radiation by dust. <u>Astrophys</u>. <u>J</u>. 167, 261.

SMITH, L.F. (1976). A review of recent observations of dust in H II regions. In <u>Solid</u> <u>state</u> <u>astrophysics</u> (eds. N.C. Wickramasinghe and D.J. Morgan). Reidel, Dordrecht.

ELMEGREEN, B.G., and LADA, C.J. (1977). Sequential formation of subgroups in OB associations. Astrophys. J. 214, 725.

KUTNER, M.L., TUCKER, K.D., CHIN, G., and THADDEUS, P. (1977). The molecular complexes in Orion. Astrophys. J. 215, 521.

WERNER, M.W., BECKLIN, E.E., and NEUGEBAUER, G. (1977). Infrared studies of star formation. Science 197, 723.

BALICK, B. (1978). Dust in planetary nebulae: observational considerations. In Planetary nebulae (ed. Y. Terzian). Reidel, Dordrecht.

MATHIS, J.S. (1978). Dust in planetary nebulae. In Planetary nebulae (ed. Y. Terzian). Reidel, Dordrecht.

CLAYTON, D.D., and WICKRAMASINGHE, N.C. (1976). On the development of infrared radiation from an expanding nova shell. Astrophys. Space Sci. 42, 463.

§9.5

ELSASSER, H., and FECHTIG, H. (eds.) (1976). Interplanetary dust and zodiacal light. Springer-Verlag, Berlin.

See §1.2.4. (Field and Cameron 1975, and McDonnell 1978).

BARLOW, M.J., and SILK, J. (1977). Graphite grain surface reactions in interstellar and protostellar environments. Astrophys. J. 215, 800.

LATTIMER, J.M., SCHRAMM, D.N., and GROSSMAN, L. (1977). Supernovae, grains and the formation of the Solar System. Nature (London) 269, 116.

————. (1978). Condensation in supernova ejecta and isotopic anomalies in meteorites. Astrophys. J. 219, 230.

10

CHEMICAL COMPOSITION

Many important phenomena involving interstellar grains, such as electromagnetic scattering or surface physics, can be discussed within a general theoretical framework or computational model; however, as we have seen many of the quantitative details are dependent upon more specific knowledge of the chemical composition. The techniques that bear on discovering the chemical composition will be presented here. Some are rather direct, such as spectral line identification, while others are more of the nature of boundary conditions, including the origin of the grains. The latter problem is not understood sufficiently to place narrow restrictions on the composition, and so is considered separately in §11. Nevertheless some of the ideas on formation are used implicitly here.

10.1. COSMIC ABUNDANCES

10.1.1. A boundary condition
Table 10.1 shows the relative importance of the most abundant elements, as determined from the solar atmosphere and primitive meteorites. The 'heavy' elements, those other than H and He, amount to only 2 per cent by mass with respect to hydrogen. As was shown in §3.1.2, the observed ratio of extinction to hydrogen column density requires that a major portion of these 'heavy' elements be in the form of grains. Any material composed of rare elements cannot make a significant contribution to the extinction. We can now be more specific.

In using grain-model computations of interstellar extinction to find the required dust density, a key parameter is β_v, the cross-section per unit volume. Elaborate computations show that for a given material (or combination of materials in a core-mantle particle) β_v is relatively insensitive to the distributions of size, shape, or orientation, provided that

TABLE 10.1
COSMIC ABUNDANCES OF IMPORTANT ELEMENTS

Element	Relative Number	Relative Mass
H	10^4	10^3
C	3.70	4.44
N	1.17	1.64
O	6.76	10.8
Mg	0.34	0.82
Si	0.32	0.90
Fe	0.26	1.46

the size parameter is chosen so that the wavelength dependence of the optical extinction is matched. Using these values of $\beta_v(V)$ and N_H/E_{B-V} from
eqn (9.2) it is straightforward to calculate the required ρ_d/ρ_H; the
results for several representative materials are given in Table 10.2. For
each element it is then possible to compute the ratio of the required abundance to the cosmic abundance. Note also that even more material would be
required to explain the continued rise of extinction in the ultraviolet.
The important conclusion is that C, N, and O are needed to explain the
large amount of extinction; while compounds involving Si, Mg, or Fe may be
present, they can account for only a fraction, at most a third, of the
extinction.

The apparent uniformity in the value of β_v for many different grain
models may be understood using a powerful and general method based on the
application of the Kramers-Kronig dispersion relations to the complex electric susceptibility χ of the interstellar medium. Combining the definition
of χ with our previous definition of the complex refractive index of the
medium in terms of the complex particle cross-section (eqns (2.1), (2.5),
and (2.6)), we identify

$$4\pi\chi = 4\pi(\chi'-i\chi'') = n_d(C_p-iC_e)/k \; . \tag{10.1}$$

The particular dispersion relation of interest is

$$\chi'(\omega_i) = 2\pi^{-1} \int_0^\infty \omega\chi''(\omega)/(\omega^2-\omega_i^2) \, d\omega \tag{10.2}$$

where ω is the angular frequency, and use has been made of $\chi'(\infty) = 0$. If
we consider $\omega_i = 0$ and use eqn (10.1) this may be rewritten

$$n_d(C_p/k)_{k=0} = 2\pi^{-1} n_d \int_0^\infty C_e \, k^{-2} \, dk \; . \tag{10.3}$$

The left-hand side is as evaluated in the long-wavelength limit. For exam-

TABLE 10.2
COSMIC ABUNDANCE RESTRICTIONS ON GRAIN COMPOSITION

Material	Formula	Λ	s_d (g cm^{-3})	$\beta_v^{-1}(V)$ (μ)	ρ_d/ρ_H ($\times 10^{-2}$)	O	C	Si	(Mg,Fe)	Fe
'ice'	$O_{0.58}C_{0.32}N_{0.1}H_{2.8}$	17	1.	0.18	0.58	0.27	0.27			
orthopyroxene	$(Mg,Fe)SiO_3$	116	3.6	0.09	1.1	0.42		2.9	3.1	
olivine	$(Mg,Fe)_2SiO_4$	172	3.8	0.09	1.0	0.34		1.8	3.9	
magnetite	Fe_3O_4	232	6.	0.05	0.88	0.23				4.4
iron	Fe	56	7.9	0.04	0.93					6.4
silicon carbide	SiC	38	3.2	0.05	0.47			0.33	3.9	
graphite	C	12	2.3	0.04	0.27		0.61			

ple for spheroids (Table 2.1) we have

$$n_d (C_p/k)_{k=0} = n_d Vf = \rho_d f/s_d \qquad\qquad (10.4)$$

where V is the volume of the grain, $f=(m_o^2-1)/\{L(m_o^2-1)+1\}$, and m_o^2 is the static dielectric constant of the grain material. The right-hand side of eqn (10.3), which involves the linear extinction coefficient $n_d C_e$, can be computed from the normalized extinction curve. Together with eqn (10.4) and eqn (9.2) we have

$$\rho_d/\rho_H \approx 3.0 \times 10^{-3} \; s_d f^{-1} \int_0^\infty \{A(\lambda)/A_V\}\lambda^2 \; d\lambda^{-1} \qquad\qquad (10.5)$$

where λ^{-1} is in μ^{-1}. Note that this estimate is completely general.

The region of the spectrum from the near-infrared to the optical contributes about 0.8 to the integral, while the ultraviolet region produces only another 0.3. This implies, incidentally, that the 'extra' mass density of grains required to explain the continued rise in extinction is somewhat smaller than that required in 'normal' grains. Unfortunately the far-infrared contribution is not well known; however, this can be circumvented by using m_o^2 appropriate for the near-infrared wavelength to which the integral extends (say 2 μ). The factor f may then be evaluated as a function of shape and of orientation of the electric vector with respect to the particle axes (different L between 0 and 1) for different m_o^2. When m_o^2 is small (<4) f is quite independent of shape when averaged over the two principal orientations; for example if $m_o^2=2$ then $f \approx 3/4$. If the particle is highly conducting (metallic) but nearly spherical (axis ratio < 5) then f may rise to 3-6; the combination of metallic composition with extreme elongation or flattening can lead to even higher f. Note, however, that the effects of large values of f would normally be offset by increases in both s_d and the infrared contribution to the integral. Thus the value of ρ_d/ρ_H is quite insensitive to the grain composition.

Evaluation of eqn (10.5) in this manner for nearly spherical particles of 'ice' ($m_o \approx 1.25$, $f \approx 0.5$, $s_d \approx 1$) gives $100\rho_d/\rho_H \approx 0.5$ for grains responsible for optical extinction. This ratio compares favourably with the value 0.6 found by the grain-modelling technique (Table 10.2). All factors considered, the other tabulated restrictions based on cosmic abundances are probably accurate to within 50 per cent.

10.1.2. Depletion of elements in the gas phase
Another clue to the grain composition is provided by observations of gas phase abundances of the 'heavy' elements. Most results are obtained by satellite observations of ultraviolet lines, optical lines being scarce

except for Ca II, Na I, Ti II, K I, and Fe I. The ultraviolet studies have
several advantages over the earlier optical work; many more lines for more
elements in various stages of ionization are available so that proper
account of the ionization conditions can be taken; also the abundance of H
I and H_2 can be measured at the same time, whereas optical work had to rely
on 21-cm measurements which did not necessarily correspond to the same distance along the line of sight. Nevertheless ratios such as Ca II/Na I or
Ti II/Ca II are reliable, and the relative depletion of these elements was
correctly inferred from the data.

    Fig. 10.1 provides a useful summary of the ample evidence for deple-
tion. The interpretation of the varying amounts of depletion from element
to element, the apparent fractionation that has taken place, will be discussed in §11.1.2 in the context of the formation and destruction of dust
particles. There are two points that can be made now in connection with
the calculations behind Table 10.2. First, strong depletion of elements

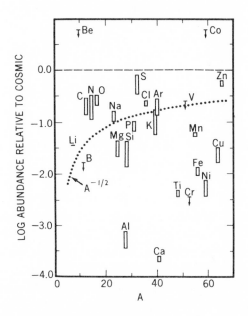

FIG. 10.1. Depletion of elements in atomic form in the H I gas
towards ζ Oph, plotted against atomic mass. Accretion on
interstellar grains would produce an $A^{-\frac{1}{2}}$ dependence for a given
grain and ionic charge. In cool interstellar clouds all elements shown, except N, O, and Ar, should be predominantly ionized. (After L. Spitzer and E.B. Jenkins, Ultraviolet studies
of the interstellar gas, A. Rev. Astron. Astrophys. 13, 133
(1975). Reproduced with permission. ©1975 by Annual Reviews
Inc.)

like Si, Mg, and Fe (and other similarly rare elements) is not unexpected, since even if they were completely depleted the compounds they form would not account for all of the extinction. However, the same is not true for C, N, and O. For illustration let us try to account for C. Table 10.2 indicates that a fraction 0.3 is needed in icy grains. Another 0.1 might be required to explain ultraviolet extinction. In gaseous form only a fraction 0.2 is seen, leaving 0.4 'missing'. CO accounts for 0.2 in clouds with $\tau > 1$; the remaining 0.2 'missing' is perhaps small enough to be explained away by the combined observational uncertainties. The situation for 'missing' N and O would be worse. The missing material might be hidden at the large-particle end of a power-law distribution of grain sizes, since there the extinction per unit mass is low, or a large number of as yet unspecified molecules might exist, perhaps accounting for the many uniden-tified far-ultraviolet interstellar lines and diffuse interstellar bands (§10.2.6). Another possibility is that much of the material is in the form of comets. Then the pattern of depletion could be more closely related to the formation of comets than of interstellar grains.

10.2. SPECTRAL FEATURES

Spectral features in the interstellar extinction curve supply information on the type of chemical compounds present in interstellar grains and on the amount of material present. Thus it is possible in principle to discover the relative contributions of various grains to interstellar extinction. Polarization measurements across these features can show whether that material contributes to interstellar polarization. Except for the diffuse bands, which remain unidentified over four decades since their discovery, the optical spectrum is notably devoid of features attributable to inter-stellar grains. Thus the extension of observations into the ultraviolet and infrared, and subsequent widespread detection of features near 2200 Å and 10 μ, was a major advance. These features have been commonly assigned to graphite and silicates respectively. The predicted 3.1-μ band of H_2O ice has also been detected, but it is neither as strong nor as ubiquitous as predicted. Several other features have been detected in certain types of source, as will be discussed.

10.2.1. The ultraviolet extinction bump at 2200 Å
As shown in Fig. 3.2 the predominant feature in the ultraviolet is a bump near 4.6 μ^{-1}. Spectrophotometric measures for 26 stars widely distributed in galactic longitude show that the central wavelength is remarkably uni-form, being λ_o = 2175 ± 30 Å. The profile is fairly symmetrical, with a FWHM of 480 Å, somewhat dependent on the adopted baseline. Thus the fea-

ture is extremely broad, with $\Delta\lambda/\lambda_o \simeq 0.2$. It was found that the colour excess $E(\lambda_o - 3300)$ is well correlated with $E(B-V)$, indicating that the grains causing this feature are interstellar, either the same as or well mixed with those responsible for optical extinction (the ratio is 5.1). Measurements of scattered light in reflection nebulae and the diffuse Galactic light indicate a decrease in the albedo at 4.6 μ^{-1}, showing that the bump is an absorption feature.

No other features have been detected in spectrophotometric scans from 2.8 to 9.0 μ^{-1}, though this range includes absorption edges expected for the ices, H_2O, CH_4, and NH_3 at 6.37, 7.41, and 5.03 μ^{-1} respectively. However, these features could be smeared out if the grains or mantles are significantly larger than the wavelength, and so firm upper limits will have to await better photometric accuracy.

Naturally the identification of the 4.6-μ^{-1} feature is of major importance. Among many interesting proposed origins are absorption edges in some silicates like orthopyroxene, colour centres in irradiated quartz, and absorption by hydrocarbons. These have not withstood detailed examination. Other possibilities are not at all exhausted, but will require extensive laboratory measurements.

The most widely debated identification is graphite; in this case the absorption is traced to plasma oscillations of the π electrons. There is other circumstantial evidence which adds to the appeal of graphite, such as the observed depletion of C in the gas. Furthermore, in the circumstellar regions surrounding late-type C-rich supergiants, the emission spectrum near 10 μ is relatively featureless (except for SiC). Graphite particles could be the dominant species condensed there and might subsequently be expelled into the interstellar medium. On the other hand, the injection of sufficient graphite is by no means certain. Furthermore, experimental evidence indicates that a graphite surface is highly reactive; virtually every O and H atom sticking to the grain will form a molecule with a C atom of the surface. If these molecules escape from the surface this is an unstable process, leading to the erosion of graphite particles on a very short time scale ($\sim t_m$).

Because the variation of the complex refractive index of graphite near the resonance is so pronounced the central wavelength of the band produced is sensitive to the grain size, the grain shape, and the presence or absence of a grain mantle. The most straightforward interpretations using Mie theory calculations show that the particles would have to be small (< 0.02 μ), round, and uncoated. However, subsequent calculations argue that some shape and orientation distributions of small spheroids might be tolerated, or that the size restriction can be relaxed to a power-law distribution of sizes, including larger grains which produce a broad back-

ground on which the narrower peak from smaller grains is superimposed. The
latter models attribute up to half of the optical extinction to graphite,
while the small particles alone would contribute much less.

In addition to the above reservations there are other grounds for
deferring acceptance. The discrepancies among different laboratory determinations of the complex refractive index are sufficient to cause large
deviations in the predicted profile, and it has been suggested that the
refractive index of the bulk material (that measured) would have to be
modified to take into account that the radius of the suggested small particles is less than the mean free path of the conduction electrons. Finally
there are lingering doubts as to the suitability of Mie theory results for
small particles, fostered by laboratory measurements of extinction by small
particles of graphite smoke. The laboratory results, theory, and interstellar extinction all disagree in the exact position of the feature, the
width of the extinction dip at 6-7 μ^{-1}, and the amount of far-ultraviolet
extinction relative to the peak extinction at 4.6 μ^{-1}. A satisfactory
resolution to these problems in still awaited.

Support for the graphite identification would be provided by detection of a second spectral feature predicted for the range 9-10 μ^{-1}. Unfortunately size and shape also affect the position and strength of this feature. The few existing observations show no clear feature; however, to be
certain it would be desirable to extend observations beyond the expected
peak to 10.5 μ^{-1}, a possibility for stars with low reddening.

It might seem improbable to find another viable solution to the origin of such a distinct and prominent feature. To dispel such a premise let
us consider one proposal; it concerns grains of primitive amorphous structure, composed of mixtures of the metal oxides MgO, CaO, SiO, SiO_2, and
FeO. Consider first MgO, without loss of generality. In bulk form MgO is
transparent to 7.65 eV (6.17 μ^{-1}), but there it exhibits a sharp exciton
peak associated with the excitation of O^{2-} ions. In small particles this
peak shifts to lower energies, being sensitive to the co-ordination number
of O^{2-} which is lower near the surface. In the limit in which the sites of
incomplete co-ordination predominate, the exciton shifts to 5.7 eV, as
judged from reflectance spectra, which is remarkably close to 4.6 μ^{-1}. If
it is a surface exciton which is to explain the interstellar peak then the
grains must be small, a $\simeq$ 50 Å. An admixture of larger grains would broaden the feature asymmetrically towards 6.2 μ^{-1}. SiO_2 is known to have a
similar surface exciton at 4.6 μ^{-1}, whereas in CaO it is at 4.45 μ^{-1}. An
amorphous mix of such oxides might therefore provide an adequate explanation.

Laboratory measurements of the extinction by an aerosol of somewhat
larger (but small, with a $\simeq$ 300 Å) particles of MgO reveal an acceptable

fit to interstellar extinction in the range 3-5 μ^{-1}, and incidentally
demonstrate the spontaneous generation of O^{2-} ions with low co-ordination
during the formation of small metal oxide particles.

Polarization measurements through the 4.6-μ^{-1} region are also awaited
with interest, as they bear directly on the shape and alignment of the
grains responsible for the feature. It seems that graphite grains should
produce little polarization. On the other hand if the mixed diatomic oxide
grains are at all associated with the interstellar 10-μ feature (possibly
caused by SiO), which is polarized, then significant polarization is to be
expected.

10.2.2. The 10-μ emission/absorption feature of silicate
We have already commented on the 10-μ feature which distinguishes the circumstellar emission around O-rich M supergiants from the relatively smooth
continuum produced by the envelopes of carbon stars. In optically thin
envelopes the feature is seen in emission. The spectrum of one well-studied star, μ Cep, is shown in Fig. 10.2. In more optically thick envelopes
around similar cool stars the feature becomes self-absorbed, as discussed

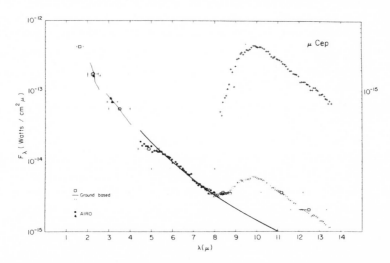

FIG. 10.2. The spectrum of μ Cep from 2 to 14 μ, incorporating
ground-based and airborne (AIRO) data. The heavy line is a fit
of a 4200 K blackbody to the narrow-band data between 5.5 and 8
μ. Using this continuum the 8-13 μ excess is derived (right-
hand scale). (From R.W. Russell, B.T. Soifer, and W.J. Forrest, Spectrophotometric observations of mu Cephei and the moon
from 4 to 8 microns, Astrophys. J. Lett. 198, L41 (1975).
University of Chicago Press. ©1975 by the American Astronomical Society. All rights reserved.)

in §7.2.2. The lower two curves (e,f) in Fig. 10.3 illustrate this. The
widespread abundance of this material is evident in emission features seen
whenever the ambient dust is heated, for example by the O stars in the Trapezium in Orion, and by the sun in comets like Comets Bennett, Bradfield,
Encke, and Kohoutek. A 10-μ absorption feature produced by cold dust is
detected against highly obscured stars like VI Cyg #12, sources embedded in
molecular clouds like the BN object in Orion (Fig. 10.3, curves a–d), and
the Galactic centre.

The central wavelength of the feature, 9.8 μ, is quite uniform in
this wide variety of sources. Similarly the profile of the absorption feature, which also shows no fine structure, matches the emission feature
well. There are of course some exceptions: the absorption feature in the
Galactic centre is somewhat narrower than usual, whereas that in the peculiar OH maser source, OH 231.8+4.2, is somewhat broader and shifted to longer wavelengths. The differences might be due to slight changes in the
composition of the material. However, the overall impression is one of
uniformity. One attractive scheme would have the grains forming in cool
circumstellar regions and pre-stellar nebulae, from which they would be
dispersed into the interstellar medium, thereby explaining their widespread
presence. However, the possibility that the grains produced in cool stellar atmospheres are different from those seen in molecular clouds should
not be overlooked.

Interstellar polarization has been observed through the 10-μ absorption feature in the BN source, other similar molecular-cloud sources, and
the Galactic centre, which shows that at least this type of grain (probably
a silicate) can be aligned. Compared with alignment in less dense regions,
magnetic (Davis-Greenstein) alignment within molecular clouds is less efficient because T_d is close to T_g and the gas density is higher. On the
other hand, if the magnetic field is amplified through flux-freezing in the
compressed cloud ($B^3 \propto n^2$), these effects may be more than offset. However,
measurements of the amount of Zeeman splitting in H I and OH lines indicate
that the field increases much less rapidly than the simple argument suggests. Therefore the problems with magnetic alignment appear to be more
severe. The exact location of the grains is uncertain. It can be noted
that if the grains are warm enough to emit far-infrared radiation (~ 100 μ),
this radiation will be polarized, with the electric vector oriented orthogonal to that at 10 μ (§6.1). An opportunity to pursue such an investigation is provided by the Kleinmann-Low nebula; it is already known that at
10 μ linear polarization is produced by transmission through aligned particles with the same orientation as seen in front of BN. If it turns out
that $T_d > T_g$, then 'inverse' Davis-Greenstein alignment should be considered
in deducing the direction of the magnetic field.

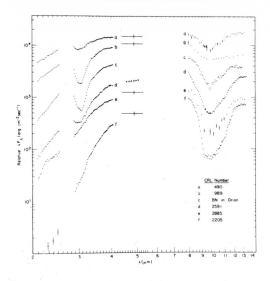

FIG. 10.3. Ground-based spectrophotometry (in the form λF_λ) of
CRL sources showing 10-μ absorption. Sources a–d are associated with dense molecular clouds, whereas e and f are cool
stars (see Fig. 7.2); near-infrared spectra are used to distinguish these classes. (From K.M. Merrill and W.A. Stein, 2–14
μm stellar spectrophotometry. III AFCRL Sky Survey objects,
Publ. astron. Soc. Pac. 88, 874 (1976). Courtesy of the
Publications of the Astronomical Society of the Pacific.)

The 10-μ feature is attributed to a Si–O stretching vibration that
occurs in all silicates. Which particular class of silicates is present
can in principle be deduced from the exact wavelength and profile, since
these vary systematically with the crystalline structure of various silicates. Present laboratory measurements follow two routes. One studies the
transmission of small-particle powders of various rocks and minerals.
Because the intrinsic band strength can be so large, the particles must be
ground down to a fine size (<0.1 μ) to avoid single-particle saturation;
saturation was a problem in early measurements. Alternatively, the bulk
refractive index can be determined from Kramers–Kronig analysis of reflectance spectra of polished samples. Scattering and extinction calculations,
as a function of shape if desirable, can then be carried out. Primarily on
the basis of wavelength mismatches of the profile, quartz and the classes
represented by the minerals olivine and orthopyroxene have been ruled out.
On the other hand the structural class of hydrous silicates, in which H_2O
is incorporated in the form of water or hydroxyls, produce spectra close to
what is observed. It is therefore interesting to note that the matrix
material of type Cl carbonaceous chondrites, the most primitive variety, is

composed primarily of such hydrous silicates (layer-lattice silicates or
phyllosilicates). Interpretation of interstellar depletion using equilibrium condensation theory (§11.1.2) suggests, however, that typical regions
of condensation may not be cool enough to produce hydrous silicates. The
amorphous compound SiO, the least complex of the silicon oxides, also produces a suitable structureless 10-μ feature.

Knowledge of the intrinsic absorption coefficient of the material is
clearly important to establishing the amount of absorbing or emitting
material present. We are therefore interested in the absorption cross-section per unit volume, called β_v previously, evaluated at the centre of the
feature. Small particles absorb in proportion to their volume, so that the
exact grain size is unimportant. (The shape can shift the wavelength of
the feature, especially when β_v is large.) The value of β_v can be calculated for spheres from

$$\beta_v = 3 \ Q_a/4a \tag{10.6}$$

and in numerical value is close to laboratory measurements of the band
strength, usually expressed in cm^{-1}. Often a mass absorption coefficient
$\beta_m = \beta_v/s_d$ (in $cm^2 \ g^{-1}$) is specified instead. The column density derived
from absorption features is simply τ/β_m. The mass of dust required to
explain an emission feature in an optically thin, constant-temperature
source can be calculated from

$$M_d = d^2 \ F(\lambda)/B(T_d,\lambda) \ \beta_m \tag{10.7}$$

where d is the distance to the source and $F(\lambda)$ is the excess flux observed
above the continuum. Eqn (10.7) is particularly useful for comparing different materials in the same source, since the ratio is independent of distance and relatively insensitive to the choice of T_d.

The spectral dependence of p/τ can be used to restrict β_v, and hence
the mineral form of the silicate, in the following way. If β_v is high,
there must be a strong variation in the complex refractive index through
the feature. It can be seen from the small-particle formulae (Table 2.1)
that the cross-section is then strongly dependent on the depolarization
factor; the effective wavelength of the feature varies with the orientation
of incident electric vector. Therefore the p/τ curve takes on a tilde-
shaped distortion. On the other hand p/τ is flat if β_v is weak. The flat
behaviour of p/τ observed in BN and the Galactic centre has been inter-
preted as showing that $\beta_v < 3 \times 10^3$ and $7 \times 10^3 \ cm^{-1}$, respectively. (Similar limits would apply to any material, silicate or not.) These stringent
limits alone are sufficient to rule out materials such as .quartz, olivine,

enstatite, and some phyllosilicates like talc. We also note that most terrestrial and lunar rocks and glasses have absorption coefficients exceeding
10^4 cm^{-1}. On the other hand some phyllosilicates including montmorillonite
(abundant in type C1 carbonaceous chondrites) and some clay minerals, as
well as actual chondritic material, have sufficiently low band strengths.
In general structurally disordered silicates (like amorphous protosilicates
produced for laboratory measurements) have suitably low β_v and a lack of
fine structure in the profile.

It is possible to use the strength of an infrared spectral feature to
predict the accompanying extinction at visual wavelengths. For example if
$\beta_v \simeq 5 \times 10^3$ cm^{-1}, then $A_v/\tau(10\ \mu) \simeq 12$ for 'normal-sized' silicate grains.
Smaller silicate grains would produce less optical extinction but would be
important in the ultraviolet. The ratio $A_v/\tau(10\ \mu)$ can rarely be determined directly, because of the required high obscuration. However, for VI
Cyg #12 a ratio 16±5 can be estimated. Observations of compact H II
regions have been used as well, A_v being estimated from the strength of the
near-infrared continuum as described in §9.4. Uncertainty in the spectral
shape of the continuum at 10 μ introduces a factor of 2 range in $\tau(10\ \mu)$,
the higher value being necessary if there is underlying 10-μ emission.
Corresponding lower limits to $A_v/\tau(10\ \mu)$ fall in the range 10–15. All of
these determinations, which are reasonably consistent, suggest that silicates contribute substantially to optical extinction, possibly to the maximum permitted by cosmic abundance constraints.

Silicates have a second major absorption feature near 20 μ, arising
from Si-O-Si bending vibrations. The exact wavelength is again sensitive
to the crystal structure, and so the position and shape of this feature are
also of potential use in the precise identification of the silicates.
Relative strengths are useful too but can be temperature sensitive. An
emission peak at 18.5 μ has already been measured spectrophotometrically in
the Trapezium (Fig. 10.4). The lack of observable fine structure and the
breadth of the feature could result from the superposition of spectra of
several substances, from mixtures of several minerals in a single grain, or
from imperfect and contaminated crystal structure. Some mixture of common
meteoritic phyllosilicates (e.g. serpentine, chlorite, and montmorillonite)
or amorphous silicates could probably explain both this and the 10-μ feature. Although pure SiO grains would have only a 10-μ feature, admixture
of FeO could produce that at 20 μ. At longer wavelengths the emission
falls off more slowly than might be expected for silicates at a single
temperature. It has been suggested that this might be an observational
problem from contamination by the wings of the cooler Kleinmann-Low nebula.
However, as we have seen (§7.2.2), a radial gradient of temperature in an
extended circumstellar region could produce the same effect; the require-

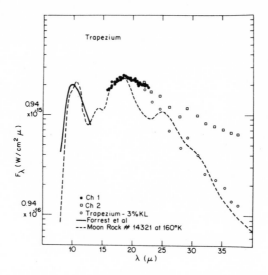

FIG. 10.4. Composite spectrum of the Trapezium infrared source
(solid curve and upper data points). Open circles are data
corrected for possible contamination by the adjacent cooler KL
nebula. A simple single-temperature silicate model (broken
curve) is also shown for comparison. (From W.J. Forrest, J.R.
Houck, and R.A. Reed, 16–40 micron spectroscopy of the Trapezium and the Kleinmann–Low nebula in Orion, Astrophys. J. Lett.
208, L133 (1976). University of Chicago Press. ©1976 by the
American Astronomical Society. All rights reserved.)

ment of cool grains would add substantially to the mass. In the BN–KL
source, which shows strong 10-µ absorption, there is only a hint of absorption near 20 µ. Nevertheless, this is quite consistent with a source that
is optically thick at 10 µ and less so ($\tau \approx 1$) at 20 µ. Observations of several other compact H II regions reveal no systematic relationship between
10 and 20-µ spectra. This may be the result of different cloud geometries.
Recall that even the combination of 10-µ absorption with 20-µ emission is
possible as predicted (Fig.7.6) and observed for circumstellar envelopes of
cool stars (e.g. CRL 2591, 2205 in Figs. 7.2 and 10.3).

10.2.3. Absorption at the 3.1-µ ice band
Water ice, supposed to be grown by accretion in the interstellar medium,
was for years considered likely to be a major constituent of interstellar
grains. It was therefore rather surprising to find no absorption at 3.1 µ,
the characteristic wavelength of the O–H stretching vibration in ice, even
in the spectra of highly reddened stars. The observational upper limit for
VI Cyg #12, which has $A_V \approx 11$, has been reduced to $\tau(3.07 \, \mu) < 0.02$. Since

for small particles of ice $\beta_m \simeq 1.4 \times 10^4$ cm^2 g^{-1}, the amount of ice is less than 1.4×10^{-6} g cm^{-2}. This is to be contrasted with about 10^{-4} g cm^{-2} expected if all the optical extinction were caused by the ice mixture in Table 10.2. Ice is not generally abundant.

However, subsequently the ice band has been discovered in many dense molecular clouds associated with regions of star formation. Four examples are shown in Fig. 10.3. The ice band is always accompanied by silicate absorption, but not vice versa. Although the feature is somewhat broader (especially noticeable in the long-wavelength wing) than calculated for small pure ice grains, there are a number of reasonable explanations. External factors, like varying shape and large size, tend to broaden strong resonances, as noted for graphite. Internal factors like imperfections in the crystal structure caused by the inclusion of impurities or other ices also shift the wavelength of the transition. Any NH_3 ice present produces absorption near 3.0 μ, in the short-wavelength wing. Finally there is the unexplored possibility that the absorption originates with the OH and H_2O of hydrous silicates. Against the latter interpretation are the similarities in silicate profile for sources with and without the ice band, and the preferential location of detectable ice bands in sources within molecular clouds. It seems quite plausible that the ice seen has formed relatively recently as a thin mantle on the silicate grains. In this case polarization at the 3.1-μ ice band should be seen in sources which show 10-μ polarization, like the BN object.

Within dense clouds the amount of ice relative to silicates is highly variable, as indicated by the range $0.2 < \tau(3.1 \ \mu)/\tau(9.8 \ \mu) < 2$. If we use 1.4×10^4 cm^{-1} and 5×10^3 cm^{-1} for the band strengths of ice and silicates respectively, then the volume ratio is 0.07–0.7. (A potential problem, masking of the 10-μ silicate feature by an absorbing mantle of ice, may not be important given this low relative abundance.) Assuming $s_d \simeq 2.5$ g cm^{-3} for silicates, the mass ratio is 0.03–0.3. Clearly even at its maximum the amount of ice falls well short of what could be allowed within cosmic abundance restraints. Why more ice is not seen is a puzzle. One possibility to be discussed in §11.2 is that simple ices formed on interstellar grains are readily transformed into more complex compounds by interaction with cosmic rays and ultraviolet photons. If this is the case the ice signature would be seen only in regions with an enhanced ice growth rate and/or shielding from energetic particles and photons; transformed mantle material would be far more common throughout the interstellar medium.

To confirm the identification of ice and learn more about its structure it is important to search for transitions at other wavelengths. Inspection of the complex refractive index of crystalline ice shows that the next two strongest resonances are at about 42 and 12.5 μ. While the imaginary part of the refractive index is actually greater at 42 μ than at 3 μ, the factor λ^{-1} in Q_a dominates to make Q_a a factor of 10 lower. At 12 μ Q_a should be about 20 times lower than at 3 μ. Present spectra of a few sources, like BN where τ(3.1 μ) ≃ 1.5, are of insufficient quality to reveal the other shallow absorption features. Polarization at 12.5 μ indicates that there is residual absorption there, but the amount appears to be too strong to be explained by ice unless the 3.1-μ band is saturated in large ice particles. (A similar effect is seen in the 10-μ polarization of the Galactic centre. It seems that the silicate absorption is either skewed towards longer wavelengths, or has an additional component near 12.5 μ, as observed in laboratory spectra of some materials.) If warm (T_d ≃ 100 K) ice grains were seen in emission the 42-μ feature should be obvious.

10.2.4. The 11.5-μ SiC emission feature

Carbon stars with circumstellar shells are characterized by an emission peak at 11.5 μ from SiC. Many IRC objects show the same feature; they are identified as heavily obscured C stars by their 2.3-μ and unidentified 3.07-μ (not ice) absorption bands which are spectral signatures of C stars. Fig. 10.5 displays the spectra of several of these sources. Broad-band spectra of some of these sources appeared in Fig. 7.2. Most of the emis-

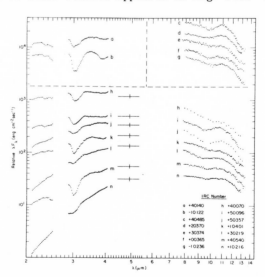

FIG. 10.5. Ground-based spectra of carbon-rich IRC sources, revealing the 11.5-μ emission feature. Sources c, i, m, and n appear in Fig. 7.2. (From K.M. Merrill and W.A. Stein, 2-14 μm spectrophotometry. II Stars from the 2 μm Infrared Sky Survey, Publ. astron. Soc. Pac. 88, 294 (1976). Courtesy of the Publications of the Astronomical Society of the Pacific.)

sion in the 8–13 μ region is in a featureless continuum on which the SiC
peak is superimposed. Another more abundant material, possibly graphite,
must be present.

The breadth of the observed peak is much larger than that calculated
for a single size and shape. However, this SiC transition is unusually
strong, with β_v reaching 2–6 × 10^5 cm^{-1} in narrow spikes for individual
grains. Therefore the central wavelength of the feature is very sensitive
to the shape- and orientation-dependent depolarization factors (see Table
2.1). Superposition of narrow components from a distribution of shapes
produces a broad emission peak spanning 10.2–12.8 μ, explaining the
observed feature. The effective β_v is reduced to 3 × 10^4 cm^{-1}.

It has been shown that another phenomenon seen in cool carbon stars,
strong opacity increasing on the short-wavelength side of 4400 Å, might be
explained by circumstellar SiC particles. (Graphite will not produce the
same effect.) However, there is no direct evidence for SiC being widely
dispersed throughout the interstellar medium.

10.2.5. Unidentified emission features

The menagerie of unidentified infrared emission features grows apace. For
example an emission peak at 11.3 μ has been discovered in the spectra of a
few objects, including the planetary nebulae NGC 7027 and BD +30 3679, the
Red Rectangle (CRL 915) and CRL 2132 (MWC 922), and the H II region NGC
7538. Thus the material is present at a variety of evolutionary stages.
Identification with magnesium carbonate $MgCO_3$ (and possibly calcium carbonate) has been explored. The fact that the feature is narrow (FWHM ≃ 0.3
μ) implies that the band strength is relatively weak, consistent with laboratory measurements of β_m ≃ 2.1 (5.5) × 10^3 cm^2 g^{-1} for these carbonates.
We note in passing that $MgCO_3$ is a reasonably volatile compound, not surviving above 350 K. Verification of this identification should be possible, since carbonates have a much stronger (β_m ≃ 9.4 (27) × 10^3 cm^2 g^{-1})
and broader feature at 6.7 μ and another feature of intermediate strength
at 28 μ.

Another narrow emission feature, somewhat weaker, has been found at
8.7 μ in the spectra of some of the same objects. It could be a strong
resonance in iron or calcium sulphate. $CaSO_4$ has β_m ≃ 1.3 × 10^4 cm^2 g^{-1}.
Estimates of the sulphate to carbonate mass ratio, using eqn (10.7), range
from 0.2 to 0.5 in the first three objects mentioned. The required abundances of both substances (∼10^{-6} $M_\odot$) are small enough that their constituent atoms will not be seriously depleted in the nebular gas. However,
other material must be present to explain the relatively continuous emission in the 8–13 μ range.

In every source in which the 11.3-μ emission appears, a narrow 3.3-μ

peak is seen; however, sometimes the 3.3-μ feature is seen where the 11.3-μ
bump is absent. A broader 3.4-μ feature always occurs with the 11.3-μ feature and so may arise in the same material. Neither near-infrared feature
breaks up into multiple components when examined at high resolution, point-
ing to a solid state rather than a molecular origin. All of the above features have been detected in some dusty galaxies (§12.2.4).

NGC 7027 has emission features at 6.2 and 7.7 μ as well; this region
of the spectrum is otherwise relatively unexplored. Carbonaceous compounds, analogous to the tarry high-molecular-weight organic material in
carbonaceous chondrites, have been mentioned as possible contributers, the
3.3 and 6.2-μ bands being associated with C-H and C=C vibrations respectively. Another suggestion is that the features are produced by fluorescence from vibrationally excited molecules in the mantles of cooler grains
at the peripheries of these sources. Not only would this provide direct
evidence for the existence of molecular mantles on composite interstellar
grains (cores of silicate material would explain the 10-μ absorption seen
in the spectra), but the chemical composition of the mantle could also be
investigated. In fact identifications of CH_4, H_2O, NH_3, and C_2H_2 have been
advanced on the basis of laboratory absorption spectra. A much fuller analysis of the implications of this hypothesis is warranted.

There is another curious phenomenon seen in one of these sources, CRL
915 (the Red Rectangle). While this may turn out to be unrelated to the
near-infrared features it seems worth mentioning here. Spectrophotometry
of the reflection nebula shows it to be unusual. It is redder than the
illuminating star HD 44179, but more importantly it shows an apparent emission feature centred at 1.5 μ^{-1} and stretching from 1.1 to 1.9 μ^{-1}. Preliminary modelling shows this peak to be consistent with an enhancement of
the scattering cross-section or albedo of the grains. However, the grain
material remains a mystery. One speculative identification of the red substance is an organic polymer-like mixture like that formed in laboratory
simulations of the synthesis of organic material in primitive planetary
atmospheres. Because of the high absolute surface brightness it has been
suggested that there might actually be fluorescence from the grains at
these wavelengths. This interesting possibility could be checked by polarization measurements, since the fluorescent emission would be unpolarized,
or at least would show a different polarization from the reflected light.

10.2.6. Diffuse interstellar bands
Characterization of the grain material from spectral features in the optical window has not been rewarding because the only possibly relevant data,
on the diffuse interstellar absorption bands, remain without satisfactory
interpretation. Association of the diffuse-band carriers with interstellar

grains has been popular, however, because of a good correlation of the strength of the bands with colour excess, and because some of the bands are quite broad and without fine structure. Any model is subject to a growing number of observational constraints. Diffuse bands are absent in circumstellar dust shells, are weak relative to extinction in dense clouds, and are relatively enhanced in low-density (sometimes high-velocity) regions. They have not been detected in reflection nebulae, although this is a rather difficult undertaking. As observations have further restricted the nature of potential host grains, support for a molecular origin has been revived. Some ideas related to a dust origin will be discussed here.

There are 39 well-established bands showing a considerable variety in depth, width, and equivalent width. The best studied are a broad feature at 4430 Å and a narrow one (in a broader trough) at 5780 Å. Two perspectives of the other lines are presented in Fig. 10.6. The strengths of the strongest features are generally very well correlated, supporting the concept of a common carrier; however, line ratios vary from star to star by up to a factor of 2. Good correlations with infrared colour excesses like E_{V-K} have been found. Though the correlation with E_{B-V} is good, it is somewhat lower; this may be related to the change of curvature in the extinction curve between the B and V passbands, which varies from star to star. A gas-phase origin of the bands is not precluded by the correlation with extinction, since the correlation may be produced spuriously by the close association of gas and dust.

A general restriction on the abundance of the carriers of the features may be obtained by combining the above-mentioned correlations with reddening with the ratio N_H/E_{B-V}. The equivalent width produced by N_a absorbers with oscillator strength f is

$$W = \pi e^2 \lambda^2 N_a f/m_e^2 c^2 . \tag{10.8}$$

For example the relation $W(4430 \text{ Å}) \simeq 2.3 E_{B-V}$ Å together with eqn (9.2) gives

$$N_a f(4430 \text{ Å})/N_H = 2 \times 10^{-9} . \tag{10.9}$$

Some perspective on the significance of this requirement may be gained by examining the cosmic abundance ratios in Table 10.1. For further comparison, the ratios of the column densities of simple abundant molecules relative to the total hydrogen column density (as observed in ζ Oph) are 4000, 20, and 7×10^{-9} for CO, CH, and CH^+ respectively.

Usually the host grains have been assumed to be those grains causing normal optical extinction. One consequence is that bands such as that at

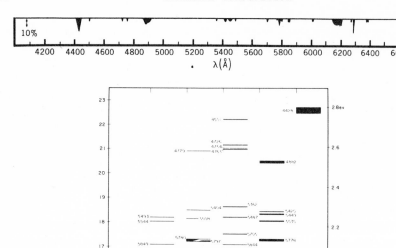

FIG. 10.6. (a) Schematic diagram of diffuse interstellar
absorption features in HD 183143, as measured by Herbig (upper
diagram). (b) A summary of many of the properties of the features arranged in the form of an energy-level diagram. The
ordinate is energy of the transition above the ground state, in
units of 10^3 cm^{-1} (left) and eV (right). The features are
grouped according to their width. The narrower lines tend to
lie at longer wavelengths. The thickness of each line indicates, on the other hand, the equivalent width. The lack of
features above 2.8 eV is real, but the situation below 1.8 eV
is unclear because adequate near-infrared data are not available. ((a) after D.G. York, Structure in the interstellar
extinction curve, Astrophys. J. 166, 65 (1971). University of
Chicago Press. ©1971 by the University of Chicago. All rights
reserved; (b) from G.H. Herbig, The diffuse interstellar bands.
IV. The region 4400–6850 Å, Astrophys. J. 196, 129 (1975).
University of Chicago Press. ©1975 by the American Astronomical Society. All rights reserved.)

4430 Å should be accompanied by a similarly broad region towards the blue
in which the extinction is lower than in the continuum; this would have the
appearance of a blue emission wing. It is now fairly clear that line profiles are symmetric and do not show such a blue emission wing. (Some nar-

row features are said to be asymmetric, being steeper on the short-wavelength side.) Thus the host grains would have to be smaller than normal.
Another related observation, that there is no polarization structure across
the bands at 4430, 5780, and 6283 Å, shows that the host grains cannot be
aligned. Since the polarizing grains are supposed to account for much of
optical extinction, this again points to another population of grains.
Small unaligned grains might be satisfactory. It is now known that diffuse
line strength does not correlate well with far-ultraviolet extinction; however, the correlation with the 2200-Å feature is good. If the host grains
were the same small particles, then the 2200-Å bump should be unpolarized,
something that should be investigated observationally.

If the carrier particles are small (<500 Å), then there are two
structural effects which can alter the positions and shapes of the diffuse
bands. First, in a small grain a large fraction of the constituent atoms,
ions, or molecules are near the surface. Consequently the local crystal
field will be quite different than in the bulk solid, and so the wavelengths of any lines from impurities, surface molecules, or defect centres
are affected. (Laboratory measurements directed toward diffuse line identification have dealt mostly with bulk samples or large particles.) The
second effect is related to the slight lattice contraction that results
from compressive surface stresses. The associated energy shift of the
transition increases inversely as the size, becoming significant below 1000
Å; a range of sizes would smear a line into a broad feature.

A final problem to be met by grain models is the apparent uniformity
in line profile and line centre from star to star. Laboratory studies have
shown how spectra of imbedded impurities depend sensitively on the interaction of the impurity with the matrix material. We have also described ways
in which grain size could modify the profile; shape and polarization can
also be important. Why the host grains are so similar would have to be
explained.

Broad-band structure in the interstellar extinction curve that might
be related to the wavelength dependence of the refractive index of the
grain material is difficult to detect. Intermediate resolution observations of the wavelength dependence of optical linear polarization have not
revealed any significant structure either. However, as this field is
largely unexplored, measurements of many more stars at high resolution seem
desirable.

10.3. CONSTRAINTS FROM CONTINUUM MEASUREMENTS

10.3.1. Extinction and polarization
Perhaps the most fundamental data to be explained by any model concern the

amount and wavelength dependence of extinction, now observed from the near
infrared to the far ultraviolet. Similarly valuable are the spectral forms
of linear and circular polarization, though they are known so far only at
optical wavelengths. Polarization involves the additional parameter,
alignment, which may be a function of size and composition. The usual
practice, which began when only optical extinction data were available, has
been to decide on the grain composition (somehow) and then to adjust the
size distribution to produce an acceptable fit. A single size served
remarkably well for the optical data alone. Size distributions decreasing
exponentially as various integer powers of the size have been used too.
While the latter help smooth out the single-particle resonances in the computed extinction curves, they have no proven basis. When observations were
extended beyond the optical the same trial-and-error technique was used,
often with several more grain components. The proliferation of substantially different models underlines the notorious non-uniqueness of this
approach.

Fitting linear polarization data suffers from the same problem. It
is known, however, that the sizes required to explain linear polarization
are approximately those needed for optical extinction. This suggests that
smaller particles (possibly of different composition) are less well
aligned, or more spherical, though a quantitative analysis of this has not
been carried out. Ultraviolet observations of polarization will be criti-
cal. With the discovery that $\lambda_c \simeq \lambda_{max}$ it was finally possible to eliminate some of the materials proposed to explain optical extinction and
polarization, such as graphite and iron; many other possibilities like ices
and silicates (or similar dielectrics) remain, however. Furthermore this
restriction on composition probably only applies to normal-sized grains, as
we have just mentioned.

Many of the large number of competing models which fit the extinction
data can be eliminated through testing their ability to satisfy other data.
The cosmic abundance restraint is an elementary example; spectral features
are also important. The use of observations of scattered light and thermal
emission will be discussed below. There will still be cause to dispute the
merits of the remaining trial-and-error models, for although the models
will be consistent with the observations they will not actually have been
predicted unambiguously.

It is therefore attractive to examine the possibility of a more
direct approach to determining the grain model. Here the experience gained
in a somewhat parallel modelling problem, the synthesis of the stellar
populations of galaxies, becomes useful. There the unknowns are the numbers of various types of star rather than the numbers of grains of different sizes, shapes, and compositions, but the same technique for determining

the numbers can be followed. It is called quadratic programming. Let us describe briefly how the problem can be set up and mention some of the preliminary results.

Ideally there are many independent observed quantities, say O_i, to be explained. The predicted quantity O_i' will be the sum of terms $n_j O_{ji}'$ representing the contributions of the n_j members of grain population j to the total. The unknowns n_j are determined by minimizing

$$\chi^2 = \sum_i (O_i - O_i')^2 / (\sigma_i O_i)^2 \tag{10.10}$$

where σ_i is the relative error of the observation O_i. The quadratic-programming approach differs from the more familiar least-squares technique by the introduction of linear inequality constraints involving the n_j. A necessary set is $n_j \geqslant 0$. The only other group used in the first application of this technique dealt with the cosmic abundance restraints element by element. Several others will be mentioned later, but first let us continue outlining the procedure that has been followed.

The O_i's used were observations of the wavelength dependence of extinction from 1 down to 0.11 μ, normalized to give the proper N_H/E_{B-V} to facilitate formulation of the abundance constraints. These observations will be referred to as E_i's. The σ_i's should reflect the relative error for each E_i. However, were it to seem desirable to give any wavelength region more weight the appropriate σ_i's could be artificially lowered. This was done for the 2200-Å extinction bump. Alternatively those E_i's could have been removed from χ^2 and been subjected to rigid constraints.

To choose the grain materials to be included some information independent of the E_i's is used. Graphite was considered because of its potential to produce the 2200-Å extinction bump. Similarly the 10-μ feature suggested the inclusion of some silicates (olivine and enstatite in this case). Selection of grain materials on the basis of spectral features of course involves some risk; for example if the 2200-Å feature is not from graphite, then the resulting models are not valid. Other materials such as SiC, Fe, and Fe_3O_4 were included on the basis of the predictions of some condensation theories of grain origin. More could have been considered. Perhaps the most serious omission was a dielectric material composed of the abundant elements H, C, N, and O. For each grain composition it is necessary to have good laboratory data on the complex refractive index.

Only spherical or infinite circular cylinder shapes were considered, for obvious computational convenience. Core-mantle particles, which might have been desirable if any icy material had been included, were also omitted for simplicity. Limiting the range of possibilities of course leaves some uncertainty in the final models. However, there is nothing in principle to prevent the inclusion of a wider variety of compositions and shapes, as long as the number and quality of independent observational data points are increased so that the extra n_j's are well determined. Possible additional data will be discussed below.

In trial-and-error modelling the size distributions and relative numbers of various grain components must be specified. The major advantage of the quadratic-programming technique is that the n_j's are determined objectively. For the purpose of computation, $n_j(a)$ was taken to be a rectangular function over a small size range $a_{j-1} < a < a_j$, with $a_j < 1.5\, a_{j-1}$. Separate n_j's for each composition and shape combination must be included (our simple subscripting here appears to allow only one combination). E'_j is the mean of the appropriate quantity over the small size range.

The derived distributions of grain size had two properties in common, a wide range in size (more than a factor of 10) and a rapid decrease with increasing size. In some of the internal size ranges n_j was found to be zero. To remove these unphysical discontinuities while maintaining the decreasing characteristic it was possible to reformulate the problem using rectangular bins covering the cumulative size ranges $0-a_j$. The resulting distributions are quite similar, resembling power laws with exponents -3.3 to -3.6 over a broad size range; several exponential-type distributions would have to be superimposed to mimic the same behaviour. Power-law distributions are attractive for two reasons: first the uniformity of the extinction curve from place to place in the Galaxy can be produced without having to rely on a common narrow range of grain sizes; second there is the hope of explaining the value of the exponent on the basis of theories of grain formation and evolution.

It was found that single-composition models fit the data very poorly. However, two-composition models, always including graphite, produced acceptable values of χ^2, while three-composition models offered still further improvement in a few cases. The silicates were not particularly favoured over SiC, Fe_3O_4, and Fe. However, the ratio $A_V/\tau(10\ \mu)$ was not included among the O_i's or even treated as a constraint, as it could be if the silicate band strength were known. If the band strength were low, much of the material besides graphite would have to be in the form of silicates.

To investigate polarization properties simple models allowing only spherical shapes for graphite particles, but using infinite cylinders for the other compositions, were constructed. Assumptions of perfect Davis-Greenstein alignment for all sizes and a viewing angle perpendicular to a uniform magnetic field were made to ensure a maximum amount of polarization. As we have seen, such perfect alignment is probably unrealistic. After fitting the models to the extinction data, using quadratic programming, the linear and circular polarization were predicted. While the

expected behaviour $\lambda_c \simeq \lambda_{max}$ was found, λ_{max} occurred at much too short a
wavelength, essentially because small aligned particles are such efficient
polarizers. To explain the observations one might assume that the smaller
particles necessary for extinction are less well aligned, or that an extra
population of larger aligned particles (contributing little to extinction)
is needed. Some objectivity can be introduced by incorporating the polarization data more directly in the quadratic-programming computation. Let
the observational data P_i in the range 0.3–1 μ have the wavelength dependence described by eqn (3.9) with $\lambda_{max} \simeq 0.55$ μ, and p_{max} chosen so that
$P_i/E_i \simeq 0.03$ when λ_i corresponds to the V filter (see eqn (3.10)). We can
conveniently write the contribution to P_i' from any subcomponent of grains
as $n_j q_j P_{ji}'$, where P_{ji}' corresponds to picket-fence alignment and $|q_j| < 1$
allows for less alignment along the lines of eqn (3.15).

Information from circular polarization data should be included also,
but there is an additional unknown, the change in the position angle of
grain alignment along the line of sight. This can be avoided by using the
semi-empirical formula for birefringence (eqn (3.16)) as the data and
allowing a small range in B_o to set up constraints. Probably only λ_i near
λ_{max} should be used because of the approximation involved, but that should
be sufficient to incorporate the important restriction $\lambda_c \simeq \lambda_{max}$.

There are two approaches to the solution. If the q_j values are specified in advance then the P_i's can be included as O_i's in the solution for
the n_j's. Suppose $q_j=0$ for graphite and $q_j=-\frac{1}{2}$ for each other composition,
as was assumed implicitly earlier. Now, however, the solution would be
quite different since the polarization must be fitted as well. The other
alternative would be to solve for the n_j's using only E_i's (and constraints) and then solve for the q_j's using the polarization data and constraints like $-\frac{1}{2}<q_j<1$. Next, new n_j's could be determined by fitting extinction and polarization simultaneously; then new q_j's could be determined,
and so on until the solution converged. This has the exciting prospect of
indicating how the degree of grain alignment varies with composition and
size.

Future applications of this technique will include other observational data, either as more O_i's or as more constraints. One obvious addition will be information on the albedo and phase function as a function of
wavelength (§10.3.2). Even if only a functional relationship between $\tilde{\omega}$ and
g, rather than independent values, can be derived observationally, the
relationship found can be usefully incorporated in the quadratic-programming technique. Additional constraints or data points could deal with the
long-wavelength emissivity, as manifested in far-infrared emission from the
Galactic plane (§10.3.3).

10.3.2. Scattered light

Radiation scattered by interstellar grains is detected in a variety of
situations, including reflection nebulae, H II regions, bright dark nebulae, and the diffuse Galactic light. Generally no information on the grain
composition can be extracted directly; however, often an attempt is made to
find the parameters g and $\tilde{\omega}$, which characterize the scattering. Since we
have just been discussing a mix composed of distinct types of particles, it
is worthwhile considering how the derived values of g and $\tilde{\omega}$ relate to individual components of the mix. For illustrative purposes imagine a simpler
mixture of normal-sized scattering particles ($\tilde{\omega} \simeq 1$, $g \simeq 0.7$) and smaller
absorbing particles ($\tilde{\omega} \simeq 0$, $g \simeq 0$) and two limiting cases, optically thin
and optically thick geometries at visual wavelengths. In the former the
presence of the absorbing particles would be irrelevant, since we obtain
information only on the grains which can scatter light efficiently; then
the observed g (and $\tilde{\omega}$) would characterize the normal-sized grains. On the
other hand in the optically thick example the absorbing grains become
important to the transfer of radiation, and so $\tilde{\omega}$ and g more closely describe the average properties of the complete mixture.

Reflection nebulae, described in §5, do not provide reliable values
of g and $\tilde{\omega}$ because of uncertainties in the geometry. However, in specific
cases some qualitative conclusions may be drawn. For example, optical
observations of the colour and polarization in the Merope Nebula indicate a
forward-throwing phase function, consistent with large dielectric particles
rather than small metallic grains. This would not rule out the presence of
the interstellar mix derived from extinction. Because the optical depth of
a nebula is ordinarily unknown, the albedo is difficult to find. Hubble's
relation, eqn (5.1), together with an estimated average optical depth ($\backsim 2$)
leads to a high albedo ($\backsim 0.8$ for reasonable g). This high value probably
describes mainly the scattering component. The interpretation of ultraviolet observations of Barnard's Loop and the Merope Nebula was mentioned
in §5.3. Finally, the high degree of polarization for the Egg Nebula and
Minkowski's Footprint seems to require small particles; recall though that
in these 'cometary' nebulae the particles probably originate locally.

Observations of scattered light in H II regions have received little
attention. This is perhaps unfortunate because thermal infrared emission,
and less directly the radio and optical emission of the gas, could be used
to specify the distribution of dust, thus eliminating some of the problems
usually associated with incomplete knowledge of the geometry of illumination. Grains in H II regions may not be typical if modified by the harsh
environment. On the other hand it would be interesting to have more observational data relevant to any such modification.

The surface brightness of a dark nebula can also be used to find $\tilde{\omega}$

and g. Monte Carlo scattering models have been constructed over a wide
range of the parameters $\tilde{\omega}$, g, and central optical depth. From these are
selected the models which best fit the observed dependence of surface
brightness on optical depth in the nebula. Only a functional relationship
between $\tilde{\omega}$ and g is found, as shown in Fig. 10.7 for the Coalsack (B filter). If the actual surface brightness profile is used then it becomes
possible to determine $\tilde{\omega}$ and g independently. This procedure has been
demonstrated for the Thumb Print Nebula, as described in §9.3. The values
g $\simeq$ $\tilde{\omega}$ $\simeq$ 0.7 (at B and V) are in good agreement with the Coalsack locus in
Fig. 10.7. It should be noted, however, that in the Thumb Print Nebula
there is evidence based on model fitting of central optical depths at various passbands for a larger than normal value of R, which suggests that the
grains may have grown in this dense cloud.

One component of the night-sky background light is the diffuse Galactic light, identified by its concentration in the Milky Way where it originates in scattering of starlight by interstellar dust. Photometric mea-

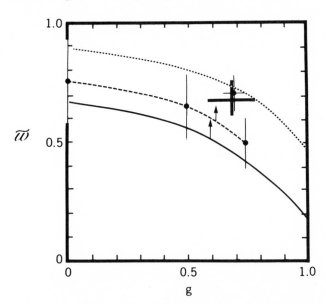

FIG. 10.7. Estimates of $\tilde{\omega}$ and g in the B passband. The lower
two loci are from diffuse Galactic light; typical uncertainties
are shown; the arrows indicate the magnitude of shift required
to take into account clumpiness in the interstellar dust layer.
The upper broken locus is from the surface brightness of the
Southern Coalsack; the typical uncertainty in $\tilde{\omega}$ is 0.1. The
thin cross is from the Thumb Print Nebula brightness distribution and the thick cross is from satellite observations of diffuse Galactic light.

surements have been made in fields avoiding stars brighter than magnitude
20, but still the light is contaminated by large contributions from airglow
and zodiacal light. The principle behind separation of these components is
their directional dependence with respect to relevant co-ordinate systems
(galactic co-ordinates, zenith distance, and ecliptic co-ordinates). Since
it is difficult to determine the small amount of diffuse Galactic light at
high galactic latitudes, values have been given with respect to a zero
point at $|b| = 27°$.

These data have been interpreted using multiple-scattering calculations within the framework of a simple plane-parallel homogenous slab model
of the Galactic disk, with the sun situated in mid-plane. The densities of
dust and stars vary with height above the mid-plane, but in the same fashion in this model. Despite the simplicity there are still two parameters
other than g and $\tilde{\omega}$ to be specified. These are the optical thickness
towards the pole, which is somewhat controversial (§9.1), and the source
function for incident starlight. A self-consistent pair for the latter was
derived by fitting the model predictions of the latitude dependence of
direct line-of-sight starlight with that computed from star counts as a
function of magnitude. Then it was possible to use the latitude dependence
of the diffuse Galactic light (with unknown zero point) to find g and $\tilde{\omega}$,
though not independently. Fig. 10.7 shows the locus of acceptable values
for the B-filter data. A subsequent interpretation of the same data using
a model with 'spiral arms gave similar results (Fig. 10.7). The upwards
arrows indicated the amount by which $\tilde{\omega}$ should be increased, about 0.07, to
allow for clumpiness in the interstellar medium. These determinations from
diffuse Galactic light agree well enough with those for individual sources,
but with the exception of the Thumb Print Nebula g and $\tilde{\omega}$ are not specified
independently. Certain limits on the albedo are found in any case.

Extension of diffuse Galactic light observations into the ultraviolet
represents a major contribution to understanding the multi-component mixtures. Satellite-based observations naturally eliminate corrections for
airglow. However, some of the fields examined by OAO-2 had stars as bright
as magnitude 14, and so the corrections for faint stars were quite important. The properties of the slab configuration were derived from star
counts as before, with one significant change: it was assumed that at the
shortest wavelengths ($\lambda < 2000$ Å) the stellar sources (being O and B stars)
would be more concentrated to the plane than the dust. This ultimately had
the effect of increasing g at these wavelengths. Since the absolute value
of the diffuse Galactic light was believed to be measured reliably at high
galactic latitudes the data were analysed using a plot of the ratio of the
diffuse light to integrated-starlight surface brightnesses against galactic
latitude. Computed curves showed that the level of this ratio in the range

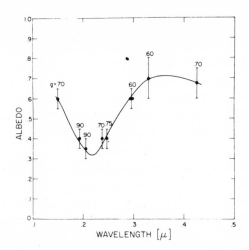

FIG. 10.8. The best estimates of albedo and g in the ultraviolet are shown as a function of wavelength. Values of g are
given above the error bars for albedo. The typical uncertainty
in g is 0.15. The smooth curve is a guide only. (From C.F.
Lillie and A.N. Witt, Ultraviolet photometry from the orbiting
astronomical observatory. XXV. Diffuse galactic light in the
1500-4200 Å region and the scattering properties of interstellar dust grains, Astrophys. J. 208, 64 (1976). University of
Chicago Press. ©1976 by the American Astronomical Society.
All rights reserved.)

$|b| > 20°$ gives $\tilde{\omega}$ independently of g, and that if g > 0.4, g is well determined by the slope of the ratio against b. Fig. 10.8 shows these results.
The errors in g are about 0.1 at $\tilde{\omega} \simeq 0.6$, increasing to 0.15 at $\tilde{\omega} \simeq 0.4$.
Note particularly the dip in albedo (and increase in g) at 2200 Å, indicat-
ing that the 2200-Å extinction bump is due to absorption. This dip has
been confirmed by independent measurements of the diffuse Galactic light by
the TD-1 satellite, and may have been seen in the surface brightness of
Barnard's Loop. The point derived for the B filter has been transferred to
Fig. 10.7, where the agreement with ground-based observations is seen to be
good. The apparent rise in albedo on the short-wavelength side of the
absorption dip is the least certain. In the same spectral range, near 1500
Å, there is little diffuse light at intermediate galactic latitudes. This
indicates that oblique scattering of light originating with OB stars concentrated in the Galactic plane is not appreciable, so that if $\tilde{\omega}$ is large,
g must be also.

    Observations of the polarization of diffuse Galactic light are not
yet advanced enough to make interpretations worthwhile. They are mentioned
here because there is some potential in this technique.

10.3.3. Thermal emission

The overall spectral distribution of emission from dust can be studied in individual sources, as discussed in §§7 and 9.4. Among the many parameters of any model that attempts to fit the spectrum is the long-wavelength emissivity of the grains. Ultimately it should be possible to separate this parameter from the others, particularly in sources which are spatially resolved. In order to assess the implications of the derived emissivity for the grain composition it will be necessary to obtain laboratory measurements of the far-infrared complex refractive index of many more materials.

It will also be important to keep in mind that the material in localized sources may not be representative of the more diffuse interstellar medium. Consequently measurements of the diffuse background far-infrared emission from dust will be interesting. As we have described in §7.1, proper intepretation will rely on independent knowledge of the interstellar radiation field responsible for heating the dust and of the spatial distribution and amount of dust. The possibility of incorporating data on thermal emission into the quadratic-programming technique has been mentioned above.

FURTHER READING

§10.1.1

CAMERON, A.G.W. (1973). Abundances of elements in the solar system. Space Sci. Rev. 15, 121.

GREENBERG, J.M., and HONG, S.S. (1974). The chemical composition and distribution of interstellar grains. In Galactic radio astronomy (eds. F.J. Kerr and S.C. Simonson). Reidel, Dordrecht.

PURCELL, E.M. (1969). On the absorption and emission of light by interstellar grains. Astrophys. J. 158, 433.

§10.1.2

MORTON, D.C. (1978). Interstellar absorption lines in the spectrum of zeta Puppis. Astrophys. J. 222, 863.

WHIPPLE, F.L. (1975). Do comets play a role in galactic chemistry and γ-ray bursts? Astron. J. 80, 525.

§10.2.1

BAR-NUN, A. (1975). Interstellar molecules: direct formation on graphite grains. Astrophys. J. 197, 341.

See §10.3.1.

DULEY, W.W., and MacLEAN, S. (1978). A possible identification of the UV extinction bump at 2175 Å. <u>Preprint</u>.

§10.2.2

MERRILL, K.M., RUSSELL, R.W., and SOIFER, B.T. (1976). Infrared observations of ices and silicates in molecular clouds. <u>Astrophys</u>. <u>J</u>. 207,
763.

NEY, E.P. (1974). Multiband photometry of Comets Kohoutek, Bennett, Bradfield, and Encke. <u>Icarus</u> 23, 551.

ZAIKOWSKI, A., KNACKE, R.F., and PORCO, C.C. (1975). On the presence of
phyllosilicate minerals in the interstellar grains. <u>Astrophys</u>. <u>Space</u>
<u>Sci</u>. 35, 97.

DAY, K.L. (1976). Temperature dependence of mid-infrared silicate absorption. <u>Astrophys</u>. <u>J</u>. <u>Lett</u>. 203, L99.

CAPPS, R.W., and KNACKE, R.F. (1976). Infrared polarization of the Galactic centre. <u>Astrophys</u>. <u>J</u>. 210, 76.

MILLAR, T.J., and DULEY, W.W. (1978). Diatomic oxide interstellar grains.
<u>Mon</u>. <u>Not</u>. <u>R</u>. <u>astron</u>. <u>Soc</u>. 183, 177.

GILLETT, F.C., JONES, T.W., MERRILL, K.M., and STEIN, W.A. (1975). Aniso-
tropy of constituents of interstellar grains. <u>Astron</u>. <u>Astrophys</u>.
45, 77.

GILLETT, F.C., FORREST, W.J., MERRILL, K.M., CAPPS, R.W., and SOIFER, B.T.
(1975). The 8–13 μ spectra of compact H II regions. <u>Astrophys</u>. <u>J</u>.
200, 609.

SARAZIN, C.L. (1978). The effect of multiple grain components on infrared
radiation transfer and the 10 micron silicate feature. <u>Astrophys</u>. <u>J</u>.
220, 165.

DYCK, H.M., and SIMON, T. (1977). Infrared observations of compact H II
regions in the spectral range 3.4–33 micrometers. <u>Astrophys</u>. <u>J</u>.
211, 421.

§10.2.3
See §10.2.2 (Merrill et al. 1976, and Gillett et al. 1975).

§10.2.4
TREFFERS, R., and COHEN, M. (1974). High resolution spectra of cool stars
in the 10- and 20-micron regions. <u>Astrophys</u>. <u>J</u>. 188, 545.

GILRA, D.P. (1973). Dust particles and molecules in the extended atmo-
spheres of carbon stars. In <u>Interstellar</u> <u>dust</u> <u>and</u> <u>related</u> <u>topics</u>
(eds. J.M. Greenberg and H.C. van de Hulst). Reidel, Dordrecht.

§10.2.5
See §7.2.1 (Merrill and Stein 1976, Part III).

PENMAN, J.M. (1976). Measurements of infrared reflectivity of astronomi-
    cally interesting non-silicates. Mon. Not. R. astron. Soc. 176,
    539.

RUSSELL, R.W., SOIFER, B.T., and WILLNER, S.P. (1978). The infrared spec-
    tra of CRL 918 and HD 44179 (CRL 915). Astrophys. J. 220, 568.

ALLAMANDOLA, L.J., and NORMAN, C.A. (1978). Infra-red emission lines from
    molecules in grain mantles. Astron. Astrophys. 63, L23.

GREENSTEIN, J.L., and OKE, J.B. (1977). An interpretation of the spectrum
    of the Red Rectangle. Publ. astron. Soc. Pac. 89, 131.

§10.2.6

SMITH, W.H., SNOW, T.P., and YORK, D.G. (1977). Comments on the origins of
    the diffuse interstellar bands. Astrophys. J. 218, 124.

DULEY, W.W. (1975). Origin of the diffuse interstellar absorption bands.
    I: Structure of interstellar grains. Astrophys. Space Sci. 36,
    345.

§10.3.1

MATHIS, J.S., RUMPL, W., and NORDSIECK, K.H. (1977). The size distribution
    of interstellar grains. Astrophys. J. 217, 425.

§10.3.2

MATTILA, K. (1971). Interpretation of the surface brightness of dark nebu-
    lae. Astron. Astrophys. 9, 53.

VAN DE HULST, H.C., and DE JONG, T. (1969). The interpretation of the dif-
    fuse Galactic radiation. Physica 41, 151.

MATHIS, J.S. (1973). The interpretation of the diffuse Galactic light.
    Astrophys. J. 186, 815.

BASTIAANSEN, P.A., and VAN DE HULST, H.C. (1977). Models for interpreting
    the diffuse Galactic light. Astron. Astrophys. 16, 1.

HENRY, R.C., ANDERSON, R., FELDMAN, P.D., and FASTIE, W.G. (1978). Far
    ultraviolet studies. III. A search for light scattered at large
    angles by dust. Astrophys. J. 222, 902.

11

EVOLUTION OF DUST GRAINS

The ultimate goal of studies of the origin and growth of grains, and coun-
tervailing destructive processes, would be to predict unambiguously the
materials present, their relative importance, and the shapes and sizes of
the particles. At present, however, these studies play an explanatory
rather than a predictive role. We are guided by many pieces of observational evidence towards plausible solutions to the problems of origin, and
subsequently discuss how the grains formed might be destroyed. A brief
overview seems warranted before we turn to a more detailed exposition.

For many years a popular explanation for the growth of interstellar
grains was the concept of accretion of the abundant elements H, C, N, and O
from the cool interstellar medium. The saturated compounds CH_4, NH_3, and
H_2O were thought to be the most likely constituents of these relatively
H-rich grains. While several modifications to this accretion hypothesis
would be necessary to satisfy present observational data, there is one outstanding reason why efforts must be made to retain it, i.e. the high value
of E_{B-V}/N_H, which requires the incorporation of a significant fraction of
C, N, and O in the grains.

A long-standing problem of any accretion process is the origin of the
seed particles on which accretion can take place. Nucleation of gas phase
atoms or molecules in low-density ($n_H \approx 10^2$ cm^{-3}), cool (T$\approx$100 K) interstellar clouds is generally regarded as unviable. Spectroscopic evidence for
the presence of refractory materials like silicates and graphite is therefore important, since these could serve as cores of core-mantle particles.
It seems likely that these cores were injected into interstellar space
after formation in a quite different environment. Indeed, the discovery of
grains in the atmospheres and circumstellar envelopes of cool evolved stars
points to one possible site; there the expulsion of grains would be driven
by radiation pressure. A related source would be grains formed in the
expanding shells of planetary nebulae, novae, and even supernovae.

A second distinct possibility would be condensation in a circumstellar nebula associated with star formation. Subsequently the nebula would
be dispersed, perhaps by a stellar wind, thus supplying new grains, or at
least grain cores, to the interstellar medium. These ideas naturally overlap substantially with discussions of the origin of our solar system
(§9.5).

Further support for the concept of refractory particles is provided
by ultraviolet studies of the interstellar gas, which reveal that the most
strongly depleted elements are those that would make up common refractory

minerals. It has been argued that pre-stellar nebulae are the principal
sites for condensation. On the other hand the carbon-rich environment
required for the production of ·graphite indicates that the observed stellar
sources are not negligible.

The distinction between volatile and refractory components of interstellar dust is also a useful concept for mechanisms of destruction. For
example, volatile mantle material, which can accrete rapidly in a molecular
cloud, is also more vulnerable to sputtering and so might be expected to be
reprocessed on a much shorter time scale than the refractories, which
require more extreme conditions for destruction just as they do for formation. Part of our task is therefore to estimate the rate at which various
phenomena take place. (In some cases, for example in star formation,
interstellar grains can be formed as well as destroyed, and so the net contribution should be considered). As we shall see the rate estimates are
plagued by uncertainties, but certain qualitative features emerge that are
likely to stand up to further scrutiny.

11.1. FORMATION OF REFRACTORY GRAINS

11.1.1. Homogeneous nucleation

The growth of microcrystals (or liquid drops) from an initially pure gas
that undergoes cooling is described by the theory of homogeneous nucleation. The aim is to find the final number N_f of monomers (gas atoms or
molecules) contained in a typical crystal or, equivalently, the number of
grains produced by condensing all the monomers. Efficient growth does not
occur until the temperature is somewhat below the condensation temperature
T_c (§11.1.2), and so if the gas is supercooled by an amount $\Delta T \ll T_c$ the
partial pressure remains enhanced by a factor $S \simeq \exp(B\Delta T/kT_c^2)$ (see eqn
(11.4)). The equilibrium concentrations of microcrystals with successively
more monomers are related by Boltzmann factors containing the appropriate
N-dependent binding energies. Thus it is possible to find the critical
value N_{cr} at which the concentration ratio at the top of the chain is
unity. Once the critical value is reached further growth to N_f is very
rapid, with the total number of particles remaining about the same. The
vapour pressure falls dramatically when most of the monomers have been condensed (eqn (11.4)). (A picture sometimes used for grain formation in
expanding atmospheres is grain nucleation without an accompanying large
depletion; the grains are then supposed to grow on a much longer time scale
by accretion of the same material. We note that this is incompatible with
formation of the original particles by homogeneous nucleation.)

The interplay between the important physical conditions (T, pressure
p, and the cooling rate) is contained in the parameter η, the product of

the rate at which monomers stick to a given site on the microcrystal and
the cooling time required for an e-fold increase in S. When $\eta \gg 1$ homogeneous nucleation will take place and

$$\Delta T/T_c \simeq 0.2 \, (\ell n \, \eta)^{-0.5} \qquad (11.1)$$
$$N_{cr} \simeq (\ell n \, \eta)^{1.5} \qquad (11.2)$$
$$N_f \simeq (\eta/\ell n \, \eta)^3 . \qquad (11.3)$$

The multiplying factors in these equations depend on poorly known thermodynamic data and could vary from compound to compound; uncertainties of a
factor of 2 in eqns (11.1) and (11.2) and of up to an order of magnitude
for N_f are possible (but there is a smaller error in the grain size). Note
that the amount of supersaturation and N_{cr} are relatively insensitive to
the concentration n_m of monomers and so within the range of applicability
of these formulae the final number density of grains has the dependence n_d
$\propto n_m^{-2}$ (dense gases produce fewer, but larger, grains).

(When the growing microcrystals are small and the binding energy B
per added monomer is large, it is natural to enquire whether deposition of
the energy B in the particle will not heat it up sufficiently that evaporation offsets sticking. For the parameter values of interest here this is
indeed the case. This situation is avoided, however, when the appropriate
monomers are diatomic or polyatomic molecules, because if a bond is broken
in the incident monomer one part can escape with much of the excess energy.
Since appropriate molecular species are present in cool circumstellar
regions, nucleation can proceed via such exchange reactions, although the
effective η may be reduced by a factor of 2 or 3.)

Values of η appropriate to various environments in which grains might
form are unfortunately uncertain, and so will be mentioned only briefly.
Consider nucleation in the photosphere of a cool ($T < T_c$) supergiant, where
η depends on p, T, the temperature scale-height, and the expansion velocity
(grains formed can drive mass loss). For carbon stars, the relevant species to be formed is graphite ($T_c < 2000$ K, see §11.1.2); η is estimated to
be $10^2 - 10^4$. (In the more dense atmosphere of a cool giant ($\eta \simeq 10^7$) larger particles could be produced, but these stars are less significant pro-
ducers of grains.) The radius of such photospheric particles ($a \simeq 3\eta/\ell n \, \eta$
Å from eqn (11.3)) would lie in the range 0.006-0.3 μ), within a factor of
3. It seems probable that the larger grains in this range would be too
large to explain the properties of the 2200-Å interstellar extinction peak.
In an oxygen-rich cool photosphere the most refractory condensates (e.g. W
in Fig. 11.1) are too rare for nucleation to occur, but for the next most
refractory materials, Al_2O_3 and Ca-Al silicates (§11.1.2), $\eta \simeq 10^2-10^4$ (and
hence the characteristic grain size is the same as for graphite).

Most of the supergiants in which circumstellar emission is seen have $T > T_c$ in the photosphere. At first sight this suggests that grains have nucleated in outflowing gas (which has escaped independently of the presence of grains). Closer scrutiny of the mass-flow rates suggests that at the radius at which the temperature falls below T_c the ambient density is so low that $\eta < 1$. Resolution of this discrepancy might lie in inhomogeneities in the flow (local enhancements of η). Alternatively, the temperature distribution over the photosphere might be patchy (related to convection cells), allowing localized nucleation of grains. (This patchiness might also help explain photometric and polarimetric fluctuations.) In any case it seems unlikely that η will be very large, and so any grains formed are small enough to become interstellar grains (particularly their cores). Size estimates are also required for models of radiative transfer in circumstellar envelopes.

Nucleation in planetary nebulae ejecta is even more poorly understood. It has been estimated that $\eta \simeq 10^2$, so that particles would be extremely small. Polarimetric evidence for small particles in compact reflection nebulae like the Egg Nebula (§5.3) is consistent with this prediction. However, clumpiness in the flow might be important (planetary nebula shells are certainly inhomogeneous at later stages) and coalescence might lead to larger sizes, but reliable quantitative results are not available. Estimates for novae give $\eta \approx 1$; this marginal value indicates that some novae might form grains, while others might not (as is observed to be the case). Supernova remnants present yet another set of physical conditions. It appears that nucleation could occur in the cooling material on the inner edge of the expanding blast wave at a time before too much mixing with swept-up interstellar material has occurred. However, details are not sufficiently developed to report here.

Similarly there has been no adequate treatment of nucleation in prestellar nebulae. Estimates of p and T have been made, but the relevant cooling time poses a problem. Of course a general feature that would favour the production of a large number of small particles, rather than the opposite, would be homogeneous nucleation at small η. Independent homogeneous nucleation of various species, a form of disequilibrium, would also produce a greater number of particles. Evidence that this might sometimes occur is provided by the magnetite crystals incorporated in the hydrated silicate matrix of types C1 and C2 carbonaceous chondrites (but consider also surface nucleation). The distinctive feature of nucleation in a prestellar nebula is that a previous generation of interstellar grains, or at least their refractory cores, might have survived, as discussed with reference to the origin of the solar system (§9.5). Then surface nucleation would dominate (§11.1.3), and the numbers and sizes of particles would not

reflect the physical conditions of the nebula in the way that homogeneous
nucleation would. In fact in the unlikely case that surface nucleation
were prevalent then the net number of particles in the interstellar medium
would not increase during star formation. (Breakup of larger particles
should be considered, however.)

11.1.2. Equilibrium condensation

Theories of condensation, as distinct from nucleation, deal with determining the composition of the condensed phase. Equilibrium condensation in a
cooling medium in which gaseous and solid phases are always in thermodynamic equilibrium is usually assumed. This is computationally straightforward since no time scales need to be introduced. The condensation can be
terminated at any temperature, depending on dispersal of the material. An
alternative called complete disequilibrium condensation, in which condensed
phases are immediately removed from further interaction, is also computationally convenient, but seems inappropriate to the study of the growth of
individual small particles. The most fully developed calculations, appropriate to an oxygen-rich primitive solar nebula cooling from a completely
vaporized state, show that equilibrium condensation played an important
role in determining the composition of the solar system. These results
serve as a guide, at least qualitatively, to what occurs in other pre-stellar nebulae and circumstellar envelopes where physical conditions are similar.

The finer details of the condensation process will not be dealt with
here. It will be sufficient to examine the characteristic condensation
temperatures at which significant fractions of specific elements are incorporated into solids. Fig. 11.1 displays some results representative of the
behaviour for equilibrium condensation at solar system abundances. Carbon-
rich condensation is quite distinct (Fig. 11.2). The fundamental difference arises because the molecule CO, being so stable, locks up all the oxygen in the carbon-rich case, and <u>vice versa</u>, and this alters what species
are available to condense. (CO is ultimately reduced by H_2 at low temperatures (<500 K), though the reaction rate could be low by that time.) The
required carbon-rich environment probably only occurs in the atmospheres
and circumstellar regions of evolved stars. Graphite would be the dominant
condensate.

Condensation into the mineral indicated occurs when the pressure and
temperature change so as to cross the curve from above. The appropriate
picture is probably cooling of the gas, either in outflowing material in a
stellar atmosphere or following an earlier hotter phase in a pre-stellar
nebula, as envisaged above. The situation can be more complex in the outer
regions of pre-stellar nebulae where the initial rise in temperature is

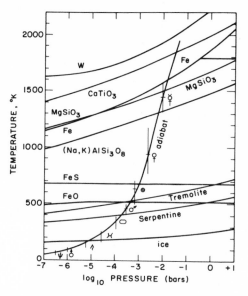

FIG. 11.1. Stability limits of condensates under equilibrium
conditions for solar abundances. W represents refractory
metals such as W and Os. $CaTiO_3$ (perovskite) represents refractory oxides and silicates, including Al_2O_3 (corundum) and
$Ca_2Al_2SiO_7$ (gehlenite). Magnesium silicates (enstatite shown)
and Fe (alloy with Ni) are seen to condense under similar conditions. Alkali aluminosilicates (alkali feldspar shown) form
at slightly lower temperatures. Sulphur is depleted as FeS
(troilite). The FeO line marks the end of oxidation of Fe;
normally FeO would be incorporated immediately into Mg-Fe silicates; if not magnetite would be formed at 400 K. The amphi-
bole tremolite and the phyllosilicate serpentine represent the
hydration of existing calcium and ferro-magnesium silicates
respectively, and thus do not imply further depletion. Water
ice, which is the principal sink for O, condenses at a relatively low temperature, and even lower temperatures are
required before solid ammonia and methane are incorporated as
hydrates. Some perspective is provided by an adiabat which
estimates the pressure-temperature profile of the pre-solar
nebula; it is labelled at the temperature points thought to
correspond to the formation conditions for various planets.
(From S.S. Barshay and J.S. Lewis, Chemistry of primitive solar
material, A. Rev. Astron. Astrophys. 14, 81 (1976). Reproduced with permission. ©1976 by Annual Reviews Inc.)

never very great. After homogeneous nucleation of a relatively refractory mineral, condensation proceeds via surface nucleation, altering the pre-existing minerals stage by stage. The amount by which a given element is depleted depends on the temperature at which condensation ceases to be efficient. Using the usual thermodynamic equation for vapour pressure and the definition of condensation temperature as that at which the vapour pressure p and undepleted partial pressure p_o match, it is straightforward to show that the depletion factor ($\xi \leqslant 1$) of the gas phase obeys ($T \leqslant T_c$)

$$\log \xi - \log(T_c/T) = \log(p/p_o) = -(B/2.3k)\ (T^{-1} - T_c^{-1}) \qquad (11.4)$$

where B is the binding energy per molecule condensed. Since refractory solids are bonded by strong valence forces, $B \approx 5$ eV (≈ 120 kcal mol^{-1}).

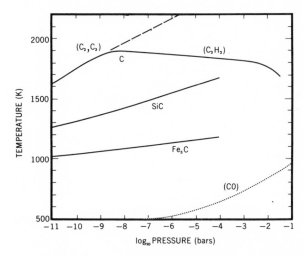

FIG. 11.2. Stability limits of condensates under equilibrium conditions for solar abundances, with the exception that C is taken to be 10 per cent more abundant than O. The condensation temperatures are not very sensitive to further enhancement of C. Above the dotted line all O is in the form of molecular CO; there is no excess O in the form of H_2O and so the condensates characteristic of the oxygen-rich case do not form. The excess C exists in molecular form as C_2 and C_3 at low pressures, condensing into graphite at the line labelled C. This line flattens at higher pressures when C_2H_2 becomes the dominant molecular species. At lower temperatures the next most abundant condensates are SiC and Fe_3C. (After E.E. Salpeter, Nucleation and growth of dust grains, Astrophys. J. 193, 579 (1974). University of Chicago Press. $^\circ$1974 by the American Astronomical Society. All rights reserved.)

Therefore $B/k \simeq 6 \times 10^4$ K $\gg T_c$ and the potential for large amounts of depletion exists.

   To help evaluate the relevance of the condensation hypothesis for grain formation the observed values of log ζ for various elements (Fig. 10.1) are replotted in Fig. 11.3 against the condensation temperature of compounds capable of depleting that element (oxygen-rich case); the result indicates that the hypothesis has some merit. For T_c larger than 700 K there is a steady decrease in log ζ, in qualitative agreement with eqn (11.4). However, a single choice of B and T does not fit the data exactly. This is not surprising since not only would B vary from compound to com-

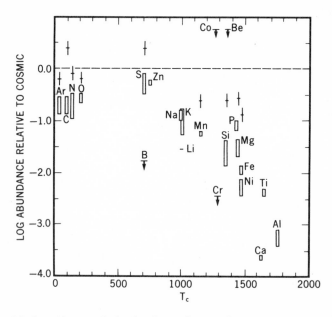

FIG. 11.3. Observed depletion of gas-phase atoms as a function of the temperature at which major condensation occurs according to equilibrium thermodynamics in a gas with cosmic (oxygen-rich) abundances at a pressure of 10^{-3} bar (10^3 dyn cm^{-2}). Labelled data are for the line of sight towards ζ Oph (as in Fig. 10.1) and are representative of cool, low-velocity interstellar material. The small crosses show the average depletions observed for intercloud, often high-velocity gas along the lines of sight to unreddened stars. (After L. Spitzer and E.B. Jenkins, Ultraviolet studies of the interstellar gas, A. Rev. Astron. Astrophys. 13, 133 (1975). Reproduced with permission. ©1975 by Annual Reviews Inc.; with data from L. Spitzer, High-velocity interstellar clouds, Comments Astrophys. 6, 177 (1976).)

pound, but also the many formation regions which have contributed to the
depletion could have different temperatures. The range of T would be 1000
- 1500 K, quite reasonable for either extended atmospheres or pre-stellar
nebulae. The decreasing reaction rates and lack of equilibrium as the
region is dispersed (as well as admixture of any unprocessed material)
would also distort the simple behaviour. Condensation sequences (and T_c)
are also dependent on p and relative abundances, and could be considerably
different for supernova dust for example.

Even so there are a number of discrepancies. That for boron might be
attributed either to adoption of an erroneous cosmic abundance or to condensation at higher T_c in some unspecified compound. The cosmic abundance
of Ar is also uncertain, so that its apparent depletion may not be real.
The observed depletion of C, N, and O (in similar amounts) does not fit
into this simple condensation picture but can be explained by accretion
onto core particles in cool interstellar clouds (§11.2).

One potentially interesting depletion anomaly is for P. Relative to
Mg and Fe in Fig. 11.3, P is somewhat less depleted, though T_c is comparable. Furthermore, P in this star, ζ Oph, is more highly depleted than
average by a factor of 2, and the effect seems likely to be further exag-
gerated by a suggested downwards revision of the crucial oscillator
strength by up to a factor of 2.3. It has also been found that the depletion of P, like that for Cl, a volatile, does not vary as widely from star
to star as do the values for Fe, Ca, and other refractory elements. The
Ca-Ti anomaly is also widespread. This makes the status of the equilibrium
condensation hypothesis, and of the related possibility that the trend with
T_c has been produced (or at least modified) by collisional destructive processes (§11.3.1) operating selectively more efficiently on more volatile
compounds, somewhat uncertain.

Some important consequences resulting from adherence to the hypothesis of equilibrium condensation can be explored. Suppose that a certain
fraction f of the interstellar medium does not show the effects of being
processed through an effective condensation environment but has a certain
amount of depletion ξ_f from another mechanism, unspecified but perhaps
related to accretion. (The metal abundance can be assumed to be the same
since enrichment of the interstellar medium proceeds on a longer time scale
than destruction of grains and mixing.) Clearly $f < \xi \xi_f^{-1}$. Substituting ξ
$\approx 10^{-3}$ (e.g. Ca, Al, Ti) and $\xi_f > 0.1$ (accretion of C,N,O), we have $f < 10^{-2}$,
which is a remarkably low fraction. Take as an extreme example the view
that all elements condensed into dust during the aftermath of a supernova
explosion in which they were produced; since their origin less than 1 per
cent of the most refractory grains can have been destroyed and left in the
interstellar gas. This seems unreasonable in the light of observational

evidence for reduced depletion due to sputtering on a time scale of less
than 10^{10} y for even the refractory particles. More generally, any effective destruction mechanism operating on a time scale comparable with that
for reprocessing the material argues against equilibrium condensation being
the dominant source of exceptional depletion; estimates indicate that suitable destructive processes do in fact exist (Fig. 11.5). A corollary is
that the observed dependence of depletion on T_c cannot be a consequence of
selective sputtering of an originally completely condensed phase. Furthermore, other mechanisms that are invoked to explain the high depletions
observed ($\xi < 10^{-3}$) must operate significantly more quickly than the known
destructive processes; in this case sputtering could contribute to the pattern of depletion by introducing a destruction rate dependent on T_c (or
more generally on the binding energy).

It is most important to note that failure to account for extreme
depletion does not rule out equilibrium condensation in circumstellar and
pre-stellar nebulae as the dominant source of interstellar refractory
grains, since that would require only $\xi \approx 10^{-1}$ say. However, an enhancement
of depletion, perhaps on these primary particles, would be necessary at
some later stage.

11.1.3. Surface nucleation

If particles are present in a cooling dense gas, surface nucleation of successive minerals is favoured over (further) homogeneous nucleation. The
number of particles therefore remains fixed and depends on η appropriate to
the original, most refractory species. (If interstellar particles survive
in a pre-stellar nebula then the number of particles will not be affected
by nucleation.) The order in which species form is given by the results of
equilibrium condensation. In a carbon-rich environment where graphite, the
dominant constituent, is the first to nucleate, the new solids are not very
important contributors to the total grain mass. However, spectroscopically
interesting materials like SiC could be produced, followed by various other
carbides, sulphides, and nitrides (Fig. 11.2). On the other hand it is at
this secondary stage that major condensation of Mg-Fe silicates occurs in
oxygen-rich environments (Fig. 11.1). The new condensates, by virtue of
their larger cosmic abundances, increase the particle radius by a factor of
2 or 3. Equilibrium calculations assume that gas and solid are completely
equilibrated. Thus adequate time for diffusion of new elements through the
grain is assumed; the basic problems are the build-up of multi-atom monomers and the rearrangement of molecular structure in the crystal. One
illustration is the sequence Fe, FeS, FeO, which ends with FeO being incorporated in magnesium silicates. Hydration of silicates is another example,
which apparently went to completion in the outer regions of the solar

nebula; however, it is unlikely to occur in circumstellar outflow, because
by the time the gas is sufficiently cool ($\sim$500 K) the density, η, and the
reaction rates are prohibitively low. (This incidentally means that the
10-μ emission feature should be ascribed to some refractory rather than
hydrated silicate.) Similarly the formation of organic molecules seems
unlikely. Generally, pure minerals are produced at high η (possibly as
layers or coatings in the case of low-abundance minerals like SiC), while
at low η an inhomogeneous mixture of various materials could coexist in the
same grain. The extreme of this is the continued growth of refractory
grains in the interstellar medium.

    An example involving surface nucleation that has been considered pos-
sible is growth of amorphous grains composed principally of MgO, FeO, and
SiO, mixed in proportion to the cosmic abundances of the metals. Such a
mixed-oxide grain, which can be considered the most primitive form of silicate, involves no chemical fractionation (its make-up is such that even
trace elements can be incorporated). If at some later stage the grain were
to be reheated and reprocessed in a pre-stellar nebula an amorphous, possibly hydrated, silicate would be produced. Spectroscopic evidence for primitive metal oxide grains was cited in §10.2.

    It has also been shown that such grains could explain the observed
pattern of depletion through a process of selective accretion. We describe
this briefly to illustrate the possible intricacies of non-equilibrium surface nucleation. One basic premise is that at the grain surface there are
defect sites containing O^-, as observed in the laboratory. Such defect
sites are produced spontaneously as the grain forms. In the interstellar
medium they would arise as the end result of absorption at the 2200-Å feature by O^{2-} ions with low co-ordination (§10.2.1). Thus the potential for
the same type of depletion exists in dense regions where the grains are
produced, and in diffuse regions later (if a grain mantle does not form).
(Surface OH^- sites can also be involved.) The rest of the scheme depends
on recombination of positive ions with these sites. Recombination has
associated with it a certain excess energy ΔE that must be transferred to
the grain if the newly formed excited molecule is to be retained. This is
possible if $\Delta E \approx 5.7$ eV, corresponding to the energy required to produce
another O^{2-} surface defect; otherwise dissociation is highly probable. The
value of ΔE depends directly on the ionization potential of the ionic species, and so some elements are preferentially incorporated while others are
not. Relative depletions predicted on the basis of this theory are in good
agreement with the observations.

    It is in principle possible to learn something of the origin of ref-
ractory seed nuclei of core-mantle particles by looking at the size of the
final particle. The argument might be developed as follows. Suppose for

simplicity that the particles are spherical and were formed in the following stages: homogeneous nucleation of 'refractory oxides', subsequent surface nucleation of 'silicates' in the same region, and accretion of an
'icy' mantle at some later stage. Ca and Al should be totally condensed
during homogeneous nucleation, but whether Mg, Fe, and Si are highly
depleted in silicates during the second stage might be questioned. An
alternative would be depletion as part of accreted 'dirty ice', although
evidence seems to be to the contrary. Consider for example the gas-phase
depletion towards unreddened stars (supposed to sample the 'intercloud medium') also shown in Fig. 11.3. Although C, N, and O are not depleted, and
the extreme depletion of Mg, Si, and Fe is removed, the latter elements are
nevertheless still significantly depleted. Consider also the ultraviolet
data showing that Si is strongly depleted in H II regions, whereas N is
not. Apparently the processes which commonly destroy 'icy' grain mantles
leave Mg, Fe, and Si appreciably depleted (though the amount varies), by
implication in more refractory grains. In fact the only cases found where
these elements and Ca are not depleted are in relatively rare high-velocity
(50-200 km s^{-1}) gas; a consistent explanation for both the high velocity
and the absence of depletion is that the grains (and gas) have been overrun
by a supernova blast wave in which destruction even of refractory grains is
possible (§11.3.1).

Returning to our discussion, the volume increase from 'refractory
oxides' to the 'silicate' stage is simply derived from the cosmic abundances and appropriate solid densities to be about 14. (This corresponds
to a size increase by a factor of 3.7.) It is also clear that if the third
stage, accretion of C, N, and O with some H, is allowed to proceed to completion, then the volume will increase by a further factor of 14. Consider
now the final size of the interstellar particles from which we hope to
deduce the size of the 'refractory oxide' particles and hence η for their
place of nucleation. Larger grains will result if the number of cores competing for the material is small, and vice versa. (This phenomenon is well
known in rain clouds.) An additional possibility to be entertained is that
only a fraction f might be able to accrete material (e.g. if temperature
fluctuations were important).

Since C, N, and O are quite highly depleted in normal clouds (Fig.
11.3), further accretion will not change the grain radius appreciably; we
adopt 0.3 μ for an 'ice' particle. Consequently the properties of the original 'refractory oxides' are as follows: radius $\sim$ 500 f$^{1/3}$ Å and number of
monomers $\sim$ 10^{7} f. These are very small particles, implying that $\eta \simeq$ 150 f
(eqn (11.3) is applied at the limit of its range of validity). A similar
analysis taking 'silicates' to be the primary particles gives $\eta \simeq$ 5000 f.
These values are comfortably close to the estimates for probable regions of

nucleation mentioned above. However, the results of this simplified derivation, with its considerable numerical uncertainty, should be regarded as
tentative. A more detailed analysis would attempt to incorporate effects
such as sputtering and grain–grain collisions (shattering and coagulation)
which can alter the grain size distribution at various stages in its development.

11.2. ACCRETION OF VOLATILE ELEMENTS

The depletion of C, N, and O relative to elements like S and Zn which have
higher T_c is difficult to explain in the context of equilibrium condensation but is quite consistent with accretion on grain surfaces in cool
interstellar clouds because the lighter elements, with higher velocities,
undergo more collisions with grains. The grain mantles might also incorporate some Ne (for which there is no depletion data), but the apparent high
depletion of Ar is inconsistent with accretion (recall Fig. 10.1).

    In normal interstellar conditions O and N are neutral whereas C
exists as C II. If accretion were on cores composed of a material like a
silicate, which has a relatively high photoelectric yield and could become
positively charged (§8.1), C depletion would be suppressed. This is not
the case (depletion in graphite grains and CO does not make up the discrepancy). A number of explanations can be offered. Possibly accretion
occurs predominantly in dense shielded regions where C is neutral, or C is
accreted in the form of CO which can be produced in gas–phase ion–molecule
reactions. Alternatively, positively charged grains may not be common if
the ultraviolet flux is diminished by shielding or the photoelectric yield
is low; in fact graphite and SiC grains and any cores coated with H, N, and
O ices might even be negatively charged. (Enhanced accretion on negatively
charged grains would be consistent with the large depletion of Li, which is
also predominantly ionized.) Any C atoms sticking to graphite cores might
migrate to a growth edge and become part of the regular crystal, thereby
greatly increasing the graphite abundance above what can be contributed by
carbon–rich regions.

    The long–standing hypothesis that grains have substantial mantles
composed of ices of C, N, and O has been confronted with the unexpected
weakness of the observed 3.1-μ band of water ice (§10.2.3). The fact that
the band is seen at all, together with its presence in molecular clouds,
motivates the search for a mechanism which modifies the chemical composition but not the elemental make–up of the mantles. One promising process
is initiated by the production of free radicals through absorption of
ultraviolet photons in the mantle. Cosmic rays might also contribute. The
end–point of subsequent recombination of the radicals is uncertain. A var-

iety of complex organic molecules could be built up by recombination of
different radicals; there are some laboratory simulation experiments supporting this scheme. Alternatively, the radical might combine with a
molecule, producing a more massive radical; this in turn could propagate by
combining with another molecule. Thus a large hybrid organic polymer-like
(tarry) substance might be grown before eventual termination by attachment
to an atom or radical. In either case significant ultraviolet processing
is expected on a time scale of about 10^6 y, short enough compared with the
depletion time scale (eqn (8.5)) to prevent a build-up of water ice.
Shielding from ultraviolet light in dense molecular clouds would limit the
conversion process (perhaps until the cloud dissipates), and so some ice
would be seen there; the weak strength, however, suggests that by the time
a dense cloud has developed, O has already been tied up in processed mantle
material and CO. This is not unexpected since radiation-processed material
should be less volatile than ice and so through selection should predominate. (CO, and similarly N_2, have high vapour pressures and so are
unlikely to condense except on the coldest grains at the centres of dense
clouds.)

     A related possibility is that very little ice forms in the first
place, particularly in dense clouds where mantle growth should involve the
accretion of molecules. It has been proposed that organic molecules might
be incorporated directly into the mantle via polymerization, H_2CO being a
favoured building block and polyoxymethylene and polysaccharides possible
end products. Again, radiation processing could lead selectively to larger, more stable species. In the absence of a clear mechanism for polymerization, spectroscopic evidence can be sought; in fact some rather sweeping
identifications with the 2200-Å, 3-μ, and 10-μ features have been advanced.
Ample motivation is undoubtedly provided by the fascinating implications
for pre-biotic evolution in the interstellar medium and the solar nebula.
However, further careful studies are required to place this hypothesis on
an acceptable foundation.

     The rate at which a mantle grows by accretion is

$$da/dt = n_g\, m_g\, v_g/4s_d \cdot \qquad\qquad\qquad (11.5)$$

For C, N, and O (with H), $n_g m_g / n_H m_H \simeq 1.7 \times 10^{-2}$ (Table 10.1) and $s_d \simeq 1$ g
cm^{-3}. If we take $T_g \simeq 100$ K then $da/dt \simeq 2.7 \times 10^{-22} n_H$ cm s^{-1}. This is
independent of grain size, and so the new size distribution is merely the
old one translated. The possible effects of this on the extinction curve
are varied. A power-law distribution over a sufficient range of sizes
bracketing the size of grain most effective in the optical would lead to
little apparent effect, whereas a bimodal distribution could produce an

enormous change as small grains become relatively more important at optical wavelengths. Note also that large grains individually deplete more material while growing. These considerations are complicated by the possibility that the sticking coefficient (set equal to unity in eqn (11.5)) varies with size; composition variations with size, temperature variations with composition and size, and the effects of temperature spikes in very small grains (§6.2) should all be considered as contributing factors.

The time scale for growth by accretion

$$t_{ac} = a/(da/dt) \simeq 10^9 \, n_H^{-1} \, y \qquad (11.6)$$

for $a \approx 0.1 \, \mu$. With the assumptions we have been making about a large fraction of C, N, and O being in this size of grain, the accretion time scale must be equivalent to t_{dep}, the time scale for an individual atom to stick to any grain surface (eqn (8.5)). We have seen that t_{dep} is probably sufficiently small to favour growth even in normal interstellar conditions. Substantial depletion of C, N, and O supports these estimates. Rapid growth in dense molecular clouds is also possible before disruptive effects of star formation take place, if there is any material left to accrete.

Evidence found for grain growth in the ρ Oph dark cloud complex has been described in §9.4. The implication here is that the gas forming the cloud was not originally too highly depleted or less probably that further growth occurred by coagulation. Net growth represents only one stage in the life of a grain, since at other times accretion is effectively offset or undone by various destructive processes.

11.3. AGENTS OF DESTRUCTION

Interstellar grains are subjected to various destructive processes which limit and reverse their growth, or even destroy them. The interplay between all the relevant phenomena in the interstellar environment is a fascinating topic, but unfortunately sufficiently complex that a definitive picture has yet to be developed. A number of plausible contributing factors will be mentioned here.

11.3.1. Collisional effects

Sputtering and grain-grain collisions in interstellar shocks are among the most potent mechanisms for grain destruction. Two sources of interstellar shocks are often considered, supernova blast waves and cloud-cloud collisions. Volatile mantle material is vulnerable in both situations, but sputtering in a blast wave is required for more refractory materials. Direct evidence for the latter is provided by high velocity components of

the interstellar gas in which the depletion of refractory elements is low
or non-existent. Estimates of the thickness of these 'clouds' are consistent with their interpretation as thin moving shocks. Intermediate levels
of depletion in intermediate-velocity gas could result from dilution of the
high-velocity gas, or renewed depletion in the post-shock material.

The effectiveness of sputtering depends on the temperature of the
shocked gas and the cooling time, and so time-dependent modelling is
required. These models can be characterized by a critical velocity (above
which destruction occurs), either a shock velocity for the blast wave or a
cloud-cloud relative velocity (about twice the shock speed). Some relevant
information on sputtering has already been reviewed in §8.1; estimates of
critical velocities for several materials appear along the abscissa of Fig.
11.4. Note how volatile mantles probably survive sputtering in low-velocity (the more frequent) cloud collisions. (It should be emphasized that
the details of the predictions are sensitive to assumptions concerning the

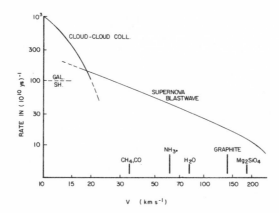

FIG. 11.4. Characteristic rate at which a given region in the
Galactic plane experiences a supernova blast wave or Galactic
spiral shock with velocity v, or a cloud-cloud collision with
relative velocity v (about twice the shock speed). Shock velocities above which significant sputtering of various compounds
will occur are estimated along the abscissa. The modelling
underlying this disarmingly simple figure relies on a complex
set of assumptions which are continually being refined as our
understanding of the interstellar medium develops. The original work should be consulted to evaluate the consequences of
adopting different parameters. (From E.E. Salpeter, Formation
and destruction of dust grains, A. Rev. Astron. Astrophys. 15,
267 (1977). Reproduced with permission. ©1977 by Annual
Reviews Inc.)

interstellar shock model and to the choice of grain parameters, and so
uncritical adoption of these results as the basis for further calculations,
without a review of the status of the.original premises, is not to be
recommended.)

Thus it is in this low-velocity limit (<10 km s^{-1}) that the consideration of grain-grain collisions becomes important. Let us estimate the
critical velocity necessary to shatter a grain by setting the relative
kinetic energy equal to the energy required for breaking a sufficient number of bonds. If the surface area of the fragments is a factor A larger
than that of the original grain and σ_b is the area per bond broken,

$$\tfrac{1}{2} m_d v_{dd}^2 = 4\pi a^2 \, A \, E_b/\sigma_b \tag{11.7}$$

where E_b is the energy per bond. Note that if A is independent of size,
then large grains (also probably more volatile) are more susceptible to
shattering. Substitution of A$\approx$10, $E_b \approx 3$ eV, $\sigma_b \simeq 10^{-15}$cm^2, and a $\simeq 10^{-5}$ cm
gives $v_{dd} \simeq 2$ km s^{-1}, illustrating the potential importance of this effect.
(If higher velocities were achieved then complete vaporization could take
place.)

It remains to suggest how grains that are originally in the same
cloud could acquire such a large relative velocity. Consider gas and
grains entering the shock. The gas is decelerated, but the grains, like
shrapnel, continue their motion. In fact if the shock is very thin they
might pass right through before being slowed down by the drag of the gas,
thus avoiding sputtering in the hot gas. However, grains charged by even a
small amount will gyrate around the magnetic field and are thereby confined
to the shock. This same effect also means that relative velocities between
the grains are comparable with their original velocities, and hence exceed
a few kilometres per second. Somewhat smaller relative velocities would
arise in any case since the slowing-down time (eqn (8.18)) is dependent on
size. The fact that drag by the gas slows the particles down does create a
problem because although this time scale exceeds the time scale for gyra-
tion (Table 8.2), it is shorter than that for grain-grain collisions. The
efficiency of shattering is reduced by the ratio of these time scales to
about 1 per cent. (Different considerations hold for high-velocity shocks,
but there sputtering is effective anyway.)

11.3.2. Regions of star formation
While there is a distinctive finality to a grain being consumed by a forming star, such an event does not upset the balance of gas and dust in the
interstellar medium. We are therefore interested in the wider effects
accompanying star formation.

The possibility of rapid accretion in molecular clouds has been noted. To this should be added the phenomenon of coagulation, which could alter the shape of the grain size distribution. Turbulent motions inferred from molecular line widths may be sufficient to drive coagulation on less than the free-fall time scale, but if velocities are too large or grains become too large (they might tend to be fluffy aggregates), then shattering will dominate. Whether some steady-state distribution is achieved is uncertain.

Interstellar grains will sublimate if raised to too high a temperature. Icy mantles are particularly vulnerable, even at 100 K, since the partial pressure of the interstellar gas is so low. Possibly ultraviolet processing will have made the mantle more cohesive. Chemical erosion, such as mentioned for graphite (§10.2.1), will also be enhanced. Higher temperatures can be encountered in more localized regions in a star-forming cloud, in which case not only will the mantle disappear but also the refractory core might be vaporized. As we have seen, new grain cores probably grow in pre-stellar nebulae when the material cools down again; furthermore as the grains are expelled on dissipation of the nebula, mantle growth in the dense environment might be favoured. Thus destroyed grains are replaced by fresh grains. However, during this reprocessing the number and size distribution of grains might have been altered.

Grains that are overrun by an H II region will suffer also. Loosely bound mantle material could be stripped away both by sputtering and photodesorption. Refractory cores and less volatile (ultraviolet-processed) mantles survive.

11.4. A STEADY STATE?

First we shall summarize the time scales on which the above-mentioned growth and destruction mechanisms take place in order to judge their relative importance and interrelationships (see Fig. 11.5).

Fig. 11.4 gives the rate at which an individual interstellar grain experiences a supernova blast wave, based on a Galactic supernova rate (0.025 y^{-1}) and a model for the expansion and decay of a supernova remnant. Lifetimes against sputtering can be read off for each type of grain material, assuming a destruction probability of unity for every occurence of a shock. Values for representative refractory cores and more volatile mantles are given in Fig. 11.5. The frequency of cloud-cloud collisions (as a function of relative velocity), as estimated from velocity studies of the interstellar gas, is also shown. From this the rate of grain-grain collisions which cause shattering may be found; recall that some adjustment for inefficiency must be made.

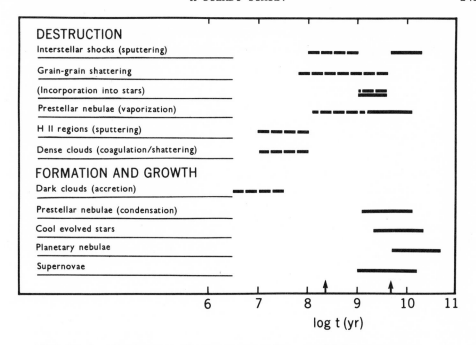

FIG. 11.5. Estimates of the time scales on which processes affecting the destruction, formation, and growth of interstellar grains occur. Time scales appropriate to refractory compounds are given by solid lines; the broken lines are for more volatile materials. For comparison the age of the sun and the period of Galactic revolution at the Galactocentric distance of the sun are indicated by arrows.

One approach to establishing time scales associated with star formation is to note that of the 4×10^9 $M_\odot$ of interstellar material in the Galaxy, an amount of 2 $M_\odot$ y^{-1} is consumed in new stars. The lifetime of a grain that is not incorporated in the star is therefore f^{-1} 2×10^9 y, where f is the fractional excess of material heated above the vaporization point in the surrounding pre-stellar nebula. For refractory materials f is small, though some estimates are as high as unity; f could be an order of magnitude or more larger for volatile mantles.

Exposure to H II regions on dissipation of the cloud and shattering and coagulation in dense clouds take in a larger fraction of the grains, and so are characterized by shorter time scales, about 10^7–10^8 y.

Accretion of mantles in dark clouds and subsequent ultraviolet processing may cycle as fast as 10^7 y. On the other hand the production of refractory grains is much slower. It seems likely that a significant proportion of the refractory material vaporized in pre-stellar nebulae is recondensed, and so any destruction in this way is compensated.

New material is added to the interstellar medium by supernovae, novae, mass loss from cool stars, and planetary nebulae. This material is not only enriched in heavy elements, but observations indicate that condensation of refractory grains occurs efficiently. However, the net effect of supernovae is not clear because the amount of refractory material sputtered by the time the remnant has slowed to 200 km s^{-1} probably exceeds the amount of new refractory elements injected; thus an evaluation of re-condensation in post-shock gas becomes crucial.

Cool supergiants and giants return mass to the interstellar medium at a rate as high as 4×10^{-10} $M_\odot$ pc^{-2} y^{-1} in the solar vicinity. Comparison of this rate with the present local surface density of interstellar material, about 3 $M_\odot$ pc^{-2}, indicates a time scale of 7×10^9 y. This is appropriate for refractory grains like silicates; the time scale for graphite would be somewhat longer, depending on the relative importance of carbon-rich mass loss. Planetary nebulae also condense grains, but the mass injection rate, estimated to be $1-2 \times 10^{-10}$ $M_\odot$ pc^{-2} y^{-1}, is smaller than for cool stars. (More mass is lost in long-term steady outflow than in planetary nebula shells.) Novae produce grains at a rate which is a few orders of magnitude slower.

For comparison the age of the solar system (4.7×10^9 y) and the period of Galactic rotation at the Galactocentric distance of the sun (2×10^8 y) are indicated in Fig. 11.5.

Present evidence is consistent with refractory cores being relatively stable, the main evolution on shorter time scales involving accreted mantles of H, C, N, and O. Whether some unique steady-state size and composition distribution is established is an open question. The considerable uniformity in the ratio of total to selective extinction (optical data) suggests that similar phenomena are taking place throughout the Galactic disk. Some variations in dust properties from place to place are evident nevertheless: there are differences in ultraviolet extinction curves, some dense clouds appear to contain larger grains than normal, and the strength of the ice band relative to the silicate feature is variable. Such variations are in a way reassuring since extreme uniformity would be difficult to explain. It seems plausible that these variations are closely related to the growth and destruction of grain mantles, and to coagulation and shattering, for which the time scales are somewhat shorter than for recycling of refractory grains. A factor that must play an important role is that everywhere the raw materials are already substantially locked up in grains. Thus if conditions favourable to runaway accretion ever arise, the growth is soon terminated by lack of material; excessive size variations are thereby avoided.

The present state of knowledge about the origin and destruction of

interstellar dust is such that more than this sketchy treatment seems
unwarranted. All of the relevant processes may not even have been identified, and those mentioned are sufficiently uncertain in so many aspects
that precise quantitative statements are not possible. Perhaps when the
details of each contributing process have been understood, a definitive
global picture with some predictive power can be developed.

FURTHER READING

§11.1.1
SALPETER, E.E. (1974). Nucleation and growth of dust grains. Astrophys.
    J. 193, 579.
DRAINE, B.T., and SALPETER, E.E. (1977). Time-dependent nucleation theory.
    J. Chem. Phys. 67, 2230.
GALLAGHER, J.S. (1977). Dust formation in novae. Astron. J. 82, 209.
FALK, S.W., LATTIMER, J.M., and MARGOLIS, S.H. (1977). Are supernovae
    sources of presolar grains? Nature (London) 270, 700.

§11.1.2
FIELD, G.B. (1975). The composition of interstellar dust. In The dusty
    universe (eds. G.B. Field and A.G.W. Cameron). Neale Watson Academic
    Publications, New York.
JURA, M., and YORK, D.G. (1978). Observations of interstellar chlorine and
    phosphorous. Astrophys. J. 219, 861.
DAY, K.L., and DONN, B. (1978). An experimental investigation of the con-
    densation of silicate grains. Astrophys. J. Lett. 222, L45.

§11.1.3
DULEY, W.W., and MILLAR, T.J. (1978). Elemental depletions in the inter-
    stellar medium. Astrophys. J. 220, 124.

§11.2
GREENBERG, J.M., and HONG, S.S. (1974). Evolutionary characteristics of a
    bimodal grain model. In H II regions and the Galactic center (ed.
    A.F.M. Moorwood). ESRO Sci. Tech. Information Branch, Noordwijk.
WICKRAMASINGHE, N.C., HOYLE, F., BROOKS, J., and SHAW, G. (1977). Prebi-
    otic polymers and infrared spectra of galactic sources. Nature (Lon-
    don) 269, 674.
BARLOW, M.J. (1978). The destruction and growth of dust grains in inter-
    stellar space. III. Surface recombination, heavy element depletion
    and mantle growth. Mon. Not. R. astron. Soc. 183, 417.

§11.3.1

SHULL, J.M. (1977). Grain disruption in interstellar hydromagnetic shocks. Astrophys. J. 215, 105. .

See §8.1 (Barlow 1978, Onaka and Kamijo 1978).

§11.4

HAGEN, W.L. (1978). The circumstellar envelopes of M giants and supergiants. Astrophys. J. Lett. 222, L37.

BURKE, J.R., and SILK, J. (1976). The dynamical interaction of a newly formed protostar with infalling matter: The origin of interstellar grains. Astrophys. J. 210, 341.

SCALO, J.M. (1977). Grain size control in dense interstellar clouds. Astron. Astrophys. 55, 253.

EXTRAGALACTIC AND INTERGALACTIC DUST

Near the turn of this century, the recognition of obscuring equatorial bands in photographs of edge-on spiral nebulae (Fig. 12.1) cast the study of dust in other galaxies in an important dual role. Not only were the dust lanes involved in the identification of spiral nebulae as extragalactic stellar systems, but they were also used to infer the existence and distribution of obscuring material in our Galaxy. Since then investigations of dust in other galaxies have continued, but not at nearly the pace of parallel studies of our Galaxy. More recently greater importance has been attached to studying the nature of extragalactic dust, as its value in understanding various galactic phenomena has become recognized. We anticipate many exciting developments on this active frontier of astronomy.

In discussing some of what is already known we shall reserve the term extragalactic dust for dust within other galaxies (§§12.1 and 12.2) to distinguish it from possible intracluster and intergalactic dust, about which even less is known (§12.3).

12.1. EXTRAGALACTIC DUST: AN OVERVIEW

Extragalactic dust is detected by the same variety of techniques employed in investigations of interstellar dust. To provide an overview before a few specific examples are presented in §12.2, the different observational approaches which have been productive will be enumerated. Photographs of course provide dramatic evidence for dust. In spiral galaxies dust is quite concentrated to the mid-plane (Fig. 12.1) and within the plane to the spiral arms (Fig. 12.2). Irregular galaxies, particularly the Irr II subclass (e.g. M 82), also have dust clouds. On the other hand elliptical galaxies are generally dust-free, just as they are lacking in interstellar gas; this makes active elliptical galaxies with dark lanes of dust (e.g. NGC 5128) all the more remarkable.

Even when individual stars are detectable in nearby galaxies an evaluation of reddening on the basis of colour measurements is difficult, because the requisite knowledge of intrinsic colours is generally lacking. Consequently an important contribution can be made by measurements of the relative intensities of emission lines, where they exist. The intensity ratio of the S II lines at 10320 and 4072 Å is particularly useful, being insensitive to physical conditions in the gas because both lines arise from the same excited electronic level. Knowledge of only a few line ratios is less than ideal, however, since the extinction curve cannot be filled in

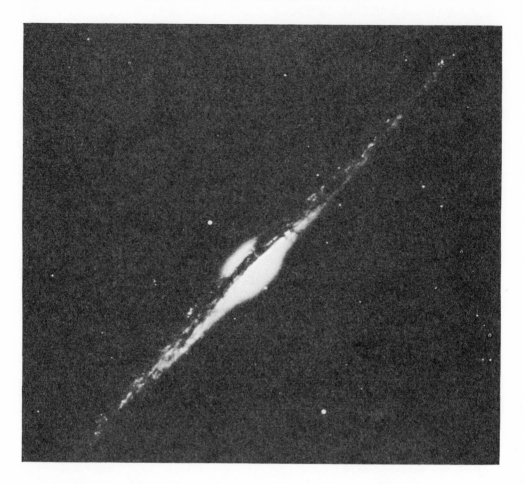

FIG 12.1. NGC 4565, a galaxy of type Sb seen edge-on in the
constellation Coma Berenices. North is up, east to left.
Scale: 1 arc min = ————. (Kitt Peak National Observatory
2.1-m photograph.)

without some assumption about the composition and size of the particles;
the situation is completely unsatisfactory if it is the nature of the
grains that is being investigated.

Detection of the linear polarization of integrated starlight transmitted through dust lanes has revealed that the grains involved have
aspherical shapes and can be aligned. Similar results have been obtained
for special cases like single stars or globular clusters in a few neigh-
bouring galaxies. Electric vectors are predominantly parallel to the equa-
tor of the galaxy, as for interstellar polarization in our Galaxy, suggesting that a similar mechanism involving a galactic magnetic field might be
responsible. Polarization patterns indicative of reflected light are also

seen, on scales ranging from the halo of a whole galaxy (e.g. M 82) down to compact regions in galactic nuclei (e.g. NGC 1068). Optical circular polarization has been detected in only one extragalactic source, the Seyfert nucleus of NGC 1068; multiple scattering, as in Galactic sources like VY CMa, seems a likely explanation.

Thermal emission from hot dust has been discovered in a variety of extragalactic sources. These range from circumstellar dust shells (e.g. Radcliffe 150 in the LMC) through large molecular clouds (e.g. in NGC 253) up to galactic nuclei whose infrared radiation dominates the bolometric luminosity of the galaxy (e.g. Markarian 231). The 10-μ 'silicate' features measured in a few objects and similar evidence at 8.7-μ represent the only spectroscopic information available on the grain composition.

Perhaps the most fundamental and important question, one that should be kept clearly in mind during all investigations, is whether extragalactic dust is at all similar to interstellar dust, in composition, shape, and size. Until this is resolved there is a danger that parallels inevitably drawn with similar phenomena involving interstellar dust will be misleading. Very little empirical evidence bearing on this question has been obtained, and so it is tempting to speculate about what kind of dust might be expected. Such speculation must have as its basis the physical conditions which affect the origin, growth, and destruction of grains; therein lies the problem, for as we have seen the environmental determinants of interstellar grains are poorly understood.

For the moment let us recall what might make extragalactic dust different. Consider nucleation. Although it might be reasoned that conditions in other spiral galaxies like are own would be on average quite similar, there are other places, for example galactic nuclei, where the apportionment of grain production among cool stellar atmospheres, planetary nebulae, pre-stellar nebulae, and supernovae might be quite different. Conditions governing nucleation in pre-stellar nebulae probably depend on the metal abundance and the mass spectrum of the stars being produced. In any case, relative elemental abundances, such as the C to O ratio and the Fe to Si-Mg ratio, determine the chemical composition of the nucleated particles. The final size of grain in turn depends on the number and size of grain cores formed and whether interstellar conditions are suitable for appreciable accretion. If all the 'heavy' elements condense, the gas-to-dust ratio will be limited by the metal abundance. In summary, reliable prediction of the properties of dust seems problematical for other spiral galaxies, let alone for unusual situations such as those encountered in active elliptical galaxies where even the presence of the gas is anomalous. Critical observations are needed. Identification of particular grain properties with specific physical conditions could lead ultimately to an understanding of the relevant processes, including those in our Galaxy.

12.2. ON THE IMPORTANCE OF EXTRAGALACTIC DUST

We have stated that the study of dust often contributes to a more complete understanding of the properties of galaxies. Diverse, though not comprehensive, illustrations of this are presented below.

12.2.1. Polarization in the Local Group

The interpretation of the linear polarization of the integrated starlight of a galaxy is somewhat ambiguous because both scattering, as in reflection nebulae, and transmission through a medium of aligned, elongated grains, as in interstellar polarization, are potential contributors. In principle it is possible to identify scattering by the distribution of the orientation of electric vectors across the surface of a galaxy (§12.2.4), but as yet such complete observational data are seldom available. If polarization by transmission were established, then the electric vector orientation would indicate the direction of alignment, from which the alignment mechanism might be determined.

The ambiguity can be avoided in the closest galaxies by observing discrete sources rather than integrated light; when the light of the discrete reddened sources is polarized, aligned grains are probably responsible. Individual bright stars have been measured in the Large and Small Magellanic Clouds (LMC and SMC respectively), whereas in M 31 globular clusters have been observed. Linear polarization was detected in each case, up to 3.0 and 2.6 per cent in the Clouds and Andromeda respectively. (Foreground polarization from dust in our Galaxy is smaller, though not always negligible.)

The first important conclusion to be drawn is that the interstellar grains there are elongated or flattened and can be aligned. As yet little is known about the physical conditions in which the alignment is taking place. However, comparison of p/E_{B-V}, which is 0.06 in the Clouds and 0.04 in M 31, with the average of 0.05 for the Galaxy shows that the resulting alignment has comparable efficiency. Observations to determine λ_{max}, which should be possible but have not been made, would provide valuable information on the characteristic size of the aligned grains.

The systematic behaviour found for the orientation of the electric vectors can be used to map out the magnetic field, assuming that there is some form of magnetic alignment. Within the LMC the polarization is strongest in large emission nebulae, in particular 30 Doradus; however, no clear relationship exists between the polarization position angles there and those in the rest of the galaxy. There is a tendency for the apparent field direction to run along the length of the prominent bar of the galaxy. Polarization vectors aligned with the dark bar in another galaxy, NGC 3718, may have a similar explanation. Potentially the most interesting discovery is that the electric vectors in the SMC are well aligned and point towards the LMC; similarly the electric vectors found for outlying stars in the LMC tend to point towards the SMC. More recently the H I bridge between the Clouds has been shown to be a part of a much longer filament called the Magellanic Stream. Thus the optical polarization measurements suggest a large-scale magnetic field oriented along the Stream. The origin of the Stream is a matter of some debate. If it results from tidal interaction of the Magellanic Clouds with our Galaxy, then the dust seen in the Stream was probably formed when the gas and dust were still a part of the Clouds.

In M 31 the preferential alignment of the electric vectors with the major axis indicates a magnetic field parallel to the galactic plane. The field topology thus shares a major characteristic with that of our Galaxy. Note also that a similar external view of our Galaxy would be expected to give the same impression as obtained for M 31. This property of alignment parallel to the plane may have wide application in the interpretation of surface polarization measurements of other galaxies, like the spiral NGC 4565. The polarization of the emission lines and strong continuum of the nuclear region of the edge-on Seyfert galaxy IC 4329A could also have such an interstellar origin. Another fascinating example is the elliptical galaxy NGC 5128, where the orientation of the surface polarization is directed along the dark equatorial band.

12.2.2. Dust as a tracer of spiral structure

Photographic surveys of the spiral structure of (nearly) face-on galaxies have shown that dark dust lanes are usually better tracers of spiral structure than bright H II regions; dust always accompanies H II regions, but not vice versa. Spiral arms outlined by dust can be followed further into the nuclear region than can their luminous counterparts. Outside the nucleus dust always appears on the inside of the luminous spiral arm. The dust arm is characteristically narrower than the luminous arm, dimensions of 100 pc and 1 kpc respectively being typical. The width of the dust lane usually increases with distance from the galactic centre.

High-resolution observations of the radio continuum emission (at 21 cm) of M 51 (NGC 5194, the Whirlpool Galaxy) when combined with optical photographs (Fig. 12.2) provide an unparalleled opportunity to examine spiral structure in the context of the density-wave theory. The principal features of the radio maps are ridges of emission outlining a two-armed spiral pattern. The radio arms are coincident with the dust arms rather than with the luminous arms (with many bright H II regions) to their out-

FIG. 12.2. NGC 5194 (M 51, the Whirlpool Galaxy), a supergiant
Sc galaxy seen face-on in the constellation Canes Venatici.
Its irregular companion is NGC 5195. North is up, east to
left. Scale: 1 arc min = ▬▬▬▬. (Kitt Peak National Observatory 4-m photograph.)

side; the radio arms also widen with distance outward along the arm.

    The pattern speed of the hypothesized density wave is less than the
speed of galactic rotation at radial distances where spiral arms are seen.
Thus interstellar gas and dust catch up to the spiral pattern and are compressed. The resulting spiral shock is the feature allowing a unified
explanation of the observations. Enhanced obscuration in the narrow

shocked region arises from the higher dust concentration, and possibly from
rapid accretion of mantles in the compressed gas. Compression of the magnetic field, which is frozen into the partially ionized gas, is responsible
for the peak in the non-thermal radio emission. The compression apparently
triggers new star formation; young stars and accompanying H II regions
appear downstream of the shocks in the luminous arms. Since spiral arms
trail, the shock and coinciding dust and radio arms are on the inside of
the luminous arms. In a few regions along the arm no dust lane is seen,
and the radio ridge is weak and lies along the luminous arm; somehow the
compression has been inhibited in these regions.

Extragalactic observations such as these are of obvious importance to
understanding similar phenomena in our Galaxy. Theoretical questions
relating to star formation in a spiral shock, and to the evolution of
interstellar clouds and of molecules and dust within them, merit continued
attention.

12.2.3. A dusty Seyfert nucleus

Early observations of the polarization of the optical continuum of the
nucleus of NGC 1068 were taken as evidence for an embedded source of synchrotron radiation. The wavelength dependence, with polarization decreasing towards the red, was attributed to increasing dilution of the non-thermal source by a normal galactic nucleus. The subsequent discovery of a
substantial infrared excess was explained naturally by a low-frequency
extension of the optical synchrotron spectrum. However, more recent observations show that large amounts of dust are responsible for both the polarization and the infrared excess.

A large amount of dust is indicated by the S II line ratio and the
Balmer decrement, from which $A(0.4 \mu) \simeq 3-5$. In compact sources much of
the scattered light is actually included in the observing aperture. When
allowance is made for this effect, using radiative-transfer models which
assume an albedo and forward-throwing phase function typical of interstellar grains in our Galaxy, the line-of-sight extinction rises to $A(0.4 \mu) \simeq$
8-16. This is consistent with $A_V > 6$ derived from the 10-μ 'silicate'
absorption feature in the infrared spectrum, again scaling according to the
behaviour found in our Galaxy. Optically thick scattering contributes to
the diffuse appearance ($\sim$1-2 arc sec) of the optical nucleus.

A dusty region such as this might be expected to give rise to the
optical polarization. This had been verified by measuring the polarization
of the strong emission lines in the nucleus; the Balmer lines are polarized
to the same degree and with the same position angle as the adjacent continuum. On the other hand the forbidden lines of oxygen are less strongly
polarized at a different position angle, implying that they arise in a dif-

ferent location, probably outside the region emitting the continuum and
Balmer lines. Such density stratification is also indicated by line-ratio
analysis of the electron density. Note that stratification affects the
interpretation of the distribution of dust in the source.

There are a number of reasons to favour scattering in an optically
thick asymmetrical dusty nucleus over transmission through aligned grains
as the explanation of the polarization. The slight variation of position
angle with wavelength is not very decisive. However, high resolution maps
of the polarization of the inner few arc seconds of the nucleus show electric vectors perpendicular to the radius vectors, a characteristic of
reflection nebulosity. Also, multiple scattering in an asymmetrical cloud
can produce the observed circular polarization, and in particular the high
ellipticity (5 per cent in the red), more readily than a transmission
model. The analogy with VY CMa has already been mentioned.

In the infrared critical new observations were the resolution of the
source at 10 μ (size ∿1 arc sec) and the discovery of a turnover in the
spectrum beyond 100 μ. An extended non-thermal source with this spectrum
seems implausible, but the features are naturally produced by a model
involving thermal re-radiation by heated dust particles; the size, somewhat
smaller than the diffuse optical nucleus, corresponds to the hot inner portion of the dusty nucleus; the spectrum results from the decreasing emissivity of the warm grains at longer wavelengths. Detailed models of the
radiative transfer have been developed with some success, although the
choice of input parameters, particularly the grain properties, is uncertain. While the radiation at wavelengths shorter than 30 μ can be
explained in a unified manner, the far-infrared radiation seems to require
additional dust, possibly in the molecular clouds observed (in CO) towards
the nucleus. Additional observations should lead to further refinements of
the basic model.

12.2.4. The polarized halo of M 82
Dust within the main body of the Irr II galaxy M 82 (NGC 3034), which is
immediately apparent from the patchy obscuration, certainly merits a separate study, but will be mentioned here only in passing. The central 300-pc
region is particularly interesting, being reminiscent of the nucleus of our
Galaxy where similar phenomena such as molecular clouds, infrared sources,
H II regions, non-thermal radio components, and star clusters are seen on a
somewhat smaller scale. Dust is especially important in converting the
energy of the excessively luminous associations of early-type stars into
powerful infrared emission. A plausible analogy might also be drawn with
the energetics of type 2 Seyfert nuclei, which are even more luminous.

Infrared spectroscopy reveals a deep 10-μ absorption, which is attributed to cool silicates along the line of sight. A similar deep depression is seen in the spectrum of the dusty Sc galaxy NGC 253, whose nuclear properties resemble those of M 82. In addition, 11.3 and 8.7-μ emission features are detected in both galaxies. As mentioned in §10.2.5, similar emission features have been discovered in Galactic sources, though not together with 10-μ absorption. The frequently associated unidentified 3.3 and 3.4-μ emission features are found in NGC 253 as well; M 82 has not yet been searched.

We turn our attention to the dust responsible for the diffuse optical halo surrounding M 82; as we shall see this dust might be quite different than the dust within the galaxy. Early interpretations emphasized the filamentary structure extending in a cone centred on the minor axis. Radial velocities of Hα emission lines in the filaments, being antisymmetric through the galactic disk, were taken to indicate outflow which was the result of an explosion in the galactic nucleus. Synchrotron radiation in the filaments was supposed to explain the optical polarization (little wavelength dependence) and the excitation of the line emission. Since that time elements of this model have been abandoned gradually, being replaced by a radically different interpretation, i.e. M 82 is embedded in a diffuse cloud of gas and dust. The observational evidence which motivated this major evolution will be outlined here.

Even at the outset scattering by dust was an attractive alternative, since polarization vectors were found to be roughly perpendicular to the direction to the nucleus. Observations have improved to the point where photographic polarization maps of the whole halo have been produced. These demonstrate that the polarization pattern is smoothly varying and characteristic of scattering by an optically thin extended cloud illuminated by the whole galactic disk, rather than by a single bright source in the nucleus. Small dust particles would reproduce the observed lack of wavelength dependence in the degree of polarization, but the simple extended-cloud models predict too much polarization in the outer halo where the average scattering angle is closer to 90°. This might be resolved by some adjustment of the geometry or inclusion of larger particles, though no detailed calculation has been presented. Analysis of the colour of the halo is also important to these discussions, as in unified models of reflection nebulae.

Filaments probably result from local enhancements in the density of scatterers ('clouds'), although possibly some are produced by illumination from shafts of light escaping from the patchy galactic disk. The excess brightness in a filament amounts to a mere 25–30 per cent and so the lack of substantial polarization changes is not unexpected. However, when the diffuse halo light is subtracted it is found that the excess filamentary

light is strongly polarized (50 per cent as compared with 20 per cent).
This can be explained by the relatively small range of scattering angles
for a filament in contrast to the large range for the halo.

The discovery that narrow Hα emission lines in the filamentary structure are polarized similarly to the continuum is consistent with the dust-
scattering model, since the galaxy provides a strong source of Hα illumination. This discovery is also of major importance in understanding the
dynamics. Recall that the early model of the explosion was based on radial
velocities of lines thought to be emitted by material flowing out along the
minor axis; although the minor axis is almost in the plane of the sky, a
small projection of the velocities on the line of sight is sufficient to
reproduce the observed radial velocities, for which a key characteristic is
antisymmetry through the galactic plane. However, when the model is modified so that filamentary Hα emission is merely emission from the disk scattered by outflowing dust, the radial velocities cannot be explained because
the line-of-sight component of the velocity which produced the antisymmetry
is completely dominated by the radial motion of the dust out from the
galaxy. An attempt has been made to explain the velocity pattern with
another simple model (involving no explosion) in which it is supposed that
the reflecting dust is a part of an intergalactic cloud in the M 81 group
of galaxies and that M 82, an interloper, is at present drifting at a velocity of 100–200 km s^{-1} through the cluster and this cloud. Relative
motion along the minor axis in the plane of the sky produces antisymmetry
in radial velocities, whereas a difference in systemic radial velocity of
the cloud and M 82 would produce merely a translation of the velocity
curve, rather like the effect of outflow or inflow.

Subsequent mapping of the 21-cm brightness distribution shows a
definite bridge of hydrogen between M 81 and M 82, with a significant
extended cloud around M 82 itself. A tidal encounter, with capture of gas
by M 82, is a possible explanation. The direction of motion is, however,
opposite to the sense of drift in the above model. Even stronger evidence
against the simple drift explanation is that although the H I cloud appears
to have close to the same systemic radial velocity as M 82, it is not in
uniform motion; instead it seems to be rolling around the major axis of the
galaxy with its angular momentum vector parallel to that of the suspected
hyperbolic orbit past M 81. A most important consequence is that dust particles sharing the rolling motion of the gas could explain the optical
radial velocities. Thus the simplest models have to be revised again, with
the addition of a component of motion thought at first to be unlikely.

Further observation will no doubt change the details, if not the
whole framework, of the interpretation of the velocity field. Other interesting questions to be addressed include the nature of the tidal encounter,

the effects of the gas on star formation activity in M 82, and the origin and nature of the dust particles. However, no further mention of these will be made here.

12.3. INTERGALACTIC DUST

Any discussion of intergalactic dust necessarily involves a considerable amount of speculation, because observational evidence is scarce and inter-galactic physical conditions are difficult to characterize. Nevertheless, some plausible arguments concerning the nature of the dust can be made, and prospects for improvement of the observational picture look promising in some areas. A question that arises immediately, on account of the low abundance of heavy elements that might be suspected, is whether grains of solid hydrogen could exist in intergalactic space. The cosmic microwave background radiation is the decisive factor. Small grains equilibrate at a temperature somewhat higher than the temperature of the blackbody radiation field since their absorptivity decreases towards longer wavelengths. Therefore a grain temperature at least as high as 2.7 K in the present epoch and larger in the past (or equivalently at large redshifts) is expected. Laboratory studies of the vapour pressure of solid hydrogen at low temperatures indicate that even at 2.7 K hydrogen grains would be unst-able. The existence of huge bodies of solid hydrogen, the size of cometary nuclei or larger, is perhaps possible; however, the extinction produced by such hypothetical bodies is unlikely to be important because the extinction efficiency, as measured by the surface area per unit mass, is so low. Direct evidence against solid H_2 grains is provided by the lack of troughs from Lyman-band absorption in the optical spectra of high-redshift quasi-stellar objects (QSO's).

The constituents of intergalactic dust are therefore restricted to heavier elements, and consequently any dust-forming material must have been processed by nuclear burning. One obvious scheme is that dust forms in galaxies from gas processed through stellar heavy-element nucleosynthesis, with subsequent ejection into intergalactic space accomplished directly by radiation pressure on the grains or indirectly as a result of collisional drag by a predominantly gaseous galactic wind. If the intergalactic dust originates in this fashion then it is tempting to carry over the character-sitics known for dust in our Galaxy to describe the properties of interga-lactic dust. However, allowance should be made for selectivity both in the ejection process and in destruction in the intergalactic environment.

The largest concentrations of dust might be expected in clusters of galaxies for two reasons: the concentration of galaxies (supposed to pro-duce the dust) is higher, and material expelled from a galaxy could remain bound to the cluster. Indirect support for this picture comes from the discovery of X-ray emission from hot gas within rich clusters such as the Perseus, Coma, and Centaurus clusters. Furthermore, the presence of 7-keV line emission from Fe XXV and Fe XXVI shows that the gas contains processed material in atomic form. (The relative abundance is reported to be 40 per cent of the value for our Galaxy.) It is not a great step further to contemplate the presence of dust, especially considering the evidence for dust in galaxies wherever there is gas. As described above, the dust surrounding M 82 might in some sense be intergalactic.

There is a potential contradiction in the above scheme connected with sputtering by hot gas, a major process by which intergalactic dust is destroyed. The gas discovered in rich clusters is sufficiently hot ($T \approx 10^7 - 10^8$ K) and dense ($n \approx 10^{-4} - 10^{-2}$ cm^{-3}) to produce substantial sputtering on a time scale of $10^7 - 10^8$ y, even for a tightly bound substance like graphite (see Fig. 8.3). This rapid destruction would effectively limit the accumulation of intergalactic dust in those clusters. Nevertheless, it has been estimated that the rate of production of dust may be large enough to achieve a steady state density corresponding to $\tau \approx 0.1$ across a rich cluster. Intracluster dust exposed to the hot gas will be heated by collisions (see eqn (6.7)) to an estimated temperature of 30 K. Searches for thermal emission in the far infrared would therefore place useful limits on the amount of dust. Optimistic predictions place the ratio of infrared to X-ray luminosity as high as 10^3 (dust is sufficiently abundant to dominate the cooling of the gas).

There are other possibilities. Dust concentrations might build up in poor clusters like the M 81 group where the intracluster gas remains in cool neutral form, or if the particles were injected into any cluster sufficiently rapidly they might escape before being destroyed, thus becoming truly intergalactic. Field galaxies could also contribute. There are obviously so many major uncertainties concerning the nature, origin, and lifetime of intergalactic dust that its very existence can be called into question. Consequently, progress in understanding intergalactic dust depends on restricting the range of acceptable speculations by obtaining direct evidence for the effects of dust.

Quite apart from their relevance to the study of the properties of the dust, observations of the amount of intergalactic obscuration are also needed for the interpretation of the Hubble diagram. If substantial obscuration were present, but neglected, distant galaxies would be taken to be too faint. Hence the derived deceleration parameter q_o would be too low. The magnitude of the error for $q_o > 0$ is related to the linear extinction coefficient τ' by

$$\Delta q_o \simeq c \, H_o^{-1} \, \tau' = 3 \, h^{-1} \, \tau' \tag{12.1}$$

where h is a dimensionless Hubble constant, $H_o/100$ km s^{-1} Mpc^{-1}, and on the right-hand side τ' is in units of Gpc^{-1}. There are other uncertainties in determining q_o, both in the observations and in the corrections to apply for galactic evolution, so that a value of $\Delta q_o \simeq 0.3$, and hence $\tau' \simeq 0.1$ h Gpc^{-1}, cannot be ruled out. This limit (with h=1 for the purposes of standardization) is given in Table. 12.1 for comparison with other results. Recall that the value of τ' for interstellar dust in our Galaxy is 2×10^6 Gpc^{-1}; this emphasizes the extreme dilution of any intergalactic dust. The mass density of the interstellar dust is about 10^{-26} g cm^{-3}. If it is supposed that the intergalactic grains produce the same extinction per unit mass, a shaky assumption, this value can be scaled to intergalactic conditions. The above limit would correspond to 5×10^{-34} g cm^{-3}, a negligible density as far as the dynamical evolution of the universe is concerned.

The search for intergalactic dust has been made by a variety of techniques. Unfortunately, none of the effects found are large or can definitely be ascribed to dust. A critical review will not be given here; an appreciation of the uncertainties involved and the assumptions made can best be obtained by reference to the original works. Instead the results are summarized in Table 12.1, with the proviso that the values of τ' are to be regarded as upper limits. Perhaps the most promising area for continued investigation is in clusters of galaxies, since gradients of colour and extinction could be established more readily than absolute amounts. The relationship to gaseous material already being discovered in clusters would

TABLE 12.1

OBSERVATIONAL LIMITS ON INTERGALACTIC DUST

τ' (Gpc^{-1})	Region	Evidence/remarks
2×10^6	Galactic plane	mass density $\sim 10^{-26}$ g cm^{-3}
500 0.3	within galaxy clusters; smoothed out (intergalactic)	lower surface density of faint galaxies towards centres of clusters of galaxies
5	plane of local supercluster	systematic changes of colour excess with 'supergalactic co-ordinates'; assumes R=3
0.1	intergalactic	limit on $\Delta q_0'$ (0.3) from Hubble diagram
0.06	intergalactic	change in mean colour of QSO's with redshift; assumes R=3
0.15	intergalactic	change in mean colour of QSO's with redshift; involves redshifting the interstellar extinction curve
0.12	intergalactic	deviations of QSO spectra from power-law form; assumes selective extinction with λ^{-1} dependence
0.05	intergalactic	systematic changes in spectra of giant elliptical galaxies with redshift; assumes selective extinction with λ^{-4} dependence

provide valuable restraints for continued theoretical investigations of intergalactic dust. In this connection far-infrared observations of X-ray emitting clusters, like the Perseus and Coma clusters, should be interesting.

In the topics encompassed in this book we have tried to place the study of dust in perspective. There is a strong interplay between attempting to define the nature of dust itself and evaluating the role of dust in the myriad of astronomical phenomena in which it is encountered. The quest for intergalactic dust illustrates this as well as any topic. To try to formulate an overall picture is necessarily an interdisciplinary endeavour. Despite the wealth of knowledge being accumulated, it can be argued that we are still very much at the stage of becoming aware of what we do not know. Thus an exhortation for continued observational and theoretical research related to the many aspects of this fascinating subject, making explicit a recurrent theme of our discussions above, seems an appropriate way to close.

FURTHER READING

§12.1
RODGERS, A.W. (1978). The stellar population and dust lane in Centaurus A
      (=NGC 5128). Astrophys. J. Lett. 219, L7.

§12.2.1
VISVANATHAN, N. (1974). Extragalactic optical polarimetry. In Planets,
      stars and nebulae studied with photopolarimetry (ed. T. Gehrels).
      University of Arizona Press, Tucson.
DAVIES, R.D., and WRIGHT, A.E. (1977). A tidal origin for the Magellanic
      Stream. Mon. Not. R. astron. Soc. 180, 71.

§12.2.2
LYNDS, B.T. (1974). An atlas of dust and H II regions in galaxies. Astro-
      phys. J. Suppl. 28, 391.
MATHEWSON, D.S., VAN DER KRUIT, P.C., and BROUW, W.N. (1972). A high reso-
      lution radio continuum survey of M 51 and NGC 5195 at 1415 MHz.
      Astron. Astrophys. 17, 468.
TOOMRE, A. (1977). Theories of spiral structure. A. Rev. Astron. Astro-
      phys. 16, 437.

§12.2.3
JONES, T.W., LEUNG, C.M., GOULD, R.J., and STEIN, W.A. (1977). Radiative

transfer in dust and the spectral flux distribution of NGC 1068.
Astrophys. J. 212, 52.

ANGEL, J.R.P., STOCKMAN, H.S., WOOLF, N.J., BEAVER, E.A., and MARTIN, P.G.
(1976). The origin of optical polarization in NGC 1068. Astrophys.
J. Lett. 206, L5.

WEEDMAN, D.W. (1977). Seyfert galaxies. A. Rev. Astron. Astrophys. 16,
69.

§12.2.4

OORT, J.H. (1977). The Galactic centre. A. Rev. Astron. Astrophys. 16,
295.

GILLETT, F.C., KLEINMANN, D.E., WRIGHT, E.L., and CAPPS, R.W. (1975).
Observations of M 82 and NGC 253 at 8-13 microns. Astrophys. J.
Lett. 198, L65.

SOLINGER, A., MORRISON, P., and MARKERT, T. (1977). M 82 sans explosion:
A galaxy drifts through dust. Astrophys. J. 211, 707.

COTTRELL, G.A. (1977). 21-cm observations of the interacting galaxies M 81
and M 82. Mon. Not. R. astron. Soc. 180, 71.

§12.3

PEEBLES, P.J.E. (1971). Physical cosmology. Princeton University Press,
Princeton.

MARGOLIS, S.N., and SCHRAMM, D.N. (1977). Dust in the universe? Astrophys. J. 214, 339.

SCHMIDT, K.-H. (1976). Existence and amount of intergalactic dust. In
Solid state astrophysics. (eds. N.C. Wickramasinghe and D.J. Morgan). Reidel, Dordrecht.

STECKER, F.W., PUGET, J.L., and FAZIO, G.G. (1977). The cosmic far-infrared background at high galactic latitudes. Astrophys. J. Lett.
214, L51.

INDEX